ORDER IN MULTIF

(T w⋯

OXFORD ARISTOTLE STUDIES

General Editors

Julia Annas and Lindsay Judson

Order in Multiplicity

HOMONYMY IN THE PHILOSOPHY OF ARISTOTLE

CHRISTOPHER SHIELDS

CLARENDON PRESS · OXFORD

OXFORD
UNIVERSITY PRESS

Great Clarendon Street, Oxford OX2 6DP

Oxford University Press is a department of the University of Oxford.
It furthers the University's objective of excellence in research, scholarship,
and education by publishing worldwide in

Oxford New York

Auckland Bangkok Buenos Aires Cape Town Chennai
Dar es Salaam Delhi Hong Kong Istanbul Karachi Kolkata
Kuala Lumpur Madrid Melbourne Mexico City Mumbai Nairobi
São Paulo Shanghai Singapore Taipei Tokyo Toronto

Oxford is a registered trade mark of Oxford University Press
in the UK and in certain other countries

Published in the United States
by Oxford University Press Inc., New York

First published 1999
First published in paperback 2002

British Library Cataloguing in Publication Data

Data available

Library of Congress Cataloging in Publication Data
Shields, Christopher John.
Order in multiplicity: homonymy in the philosophy of Aristotle/
by Christopher Shields.
(Oxford Aristotle studies)
Includes bibliographical references and index.
1. Aristotle—Contributions in methodology. 2. Methodology—
History. I. Title. II. Series.
B491.M45S55 1998 185—dc21 98-27571
ISBN 0-19-823715-4 (hbk.)
ISBN 0-19-925307-2 (pbk.)

3 5 7 9 10 8 6 4 2

Typeset by Invisible Ink
Printed in Great Britain
on acid-free paper by
Biddles Ltd, Guildford & King's Lynn

To K

Acknowledgements

I began preliminary research for this book while an Alexander von Humboldt-Stiftung Research Fellow, affiliated with the Seminar für Klassische Philologie, Johannes Gutenberg-Universität, Mainz. I thank the AvH Foundation for its generous support, and my hosts in Mainz, Wolfgang Bernard and Arbogast Schmitt, for their warm intellectual and personal hospitality. My research there resulted in an article published in *Archiv für Geschichte der Philosophie* in 1993, some portions of which provide a basis for the discussions of Chapter 5. I thank the editors of *AGP* for permission to use material derived from that article. I also thank Peter Momtchiloff at Oxford University Press for his support and excellent advice, together with Angela Blackburn and her skilled staff at Invisible Ink for their highly professional assistance in preparing the typescript for publication.

I composed most of the first draft of this book in 1992–3, in Oxford, while a Visiting Fellow of Corpus Christi College. The President and Fellows of that college provided an exceptional environment for sustained research and productive exchange. I thank them all, but most especially Christopher Taylor, Ewen Bowie, and Ian Bostridge. Some of my first formed thoughts on these matters were presented that same year to a seminar led by David Charles and Stephen Everson at Balliol. Their acute objections pertaining especially to Aristotle's conception of signification helped me to filter out some unhelpful matter from the early chapters.

Similar thanks are due to two sets of graduate students, one at the University of Colorado at Boulder and the other at Stanford University, who were kind enough to discuss the main themes of this book in two related seminars. I thank them especially for insisting that Aristotle's views be made accessible to scrutiny. I also thank individually William Simpson, Ellen Wagner, Richard Geenen, Richard Cameron, Michael Peirce, and Paul Studtmann, all students from whom I have learnt more than I taught.

Because I have had the good fortune to present the materials of individual chapters to audiences at a host of leading institutions and learned societies, I have enjoyed the judicious criticisms of a now uncountable number of auditors. I can still discern the influence of: Robert Bolton, Myles Burnyeat, Daniel Devereaux, Gail Fine, Güven Güzeldere, Norman Kretzmann, Frank Lewis, Gareth Matthews, and Fred Miller.

Through less formal settings, I have enjoyed stimulating exchanges on these matters with: James Anderson, Gabriela Carone, John Fisher, David

Horner, Dale Jamieson, Robert Kibbee, Scott MacDonald, Phillip Mitsis, Wes Morriston, James Nickel, Graham Oddie, Keith Quillen, Rex Welshon, and Edward Zalta. Each of them has given me something significant.

For penetrating criticisms of the entire penultimate draft, I express my deep gratitude to Lindsay Judson. His comments, both textually informed and philosophically adroit, delivered me from more errors than I care to recall.

I also welcome the opportunity to thank John Ackrill. Some years ago he very kindly—and very patiently—demonstrated to me how one ought to read a text of Aristotle. He has also demonstrated to everyone in his clear and accurate writings how a sympathetically critical engagement with an ancient author can produce new and arresting forms of philosophical insight. I have never failed to learn something of value from reading his work; and I have never been misled by looking to him as a model.

My deepest intellectual debts are largely unrecoverable to me. Terence Irwin first awakened my interest in many of these issues, not least through his intricately challenging writings on homonymy and signification in Aristotle. More proximately, he provided me detailed criticisms of a draft of this manuscript. These criticisms demonstrated abundantly that I had not always met the challenges implicit in his work; and they focused my attention on several consequential issues that had eluded me altogether. I thank him.

A different sort of debt is still more irretrievable to me. I can at least, though, dedicate this book to K, whose graceful love I have cherished all of my adult life.

Contents

Abbreviations

WORKS OF ARISTOTLE

An.	*De Anima*
An. Post.	*Posterior Analytics*
An. Pr.	*Prior Analytics*
Cael.	*De Caelo*
Cat.	*Categories*
EE	*Eudemian Ethics*
EN	*Nicomachean Ethics*
GA	*Generation of Animals*
GC	*De Generatione et Corruptione*
HA	*Historia Animalium*
Int.	*De Intepretatione*
Met.	*Metaphysics*
Meteor.	*Meteorologica*
PA	*Parts of Animals*
Phys.	*Physics*
PN	*Parva Naturalia*
Poet.	*Poetics*
Pol.	*Politics*
Rhet.	*Rhetoric*
Soph. El.	*Sophistici Elenchi*
Top.	*Topics*

WORKS OF PLATO

Crat.	*Cratylus*
Parm.	*Parmenides*
Phaed.	*Phaedo*
Phaedr.	*Phaedrus*
Phil.	*Philebus*
Pol.	*Statesman*
Prot.	*Protagorus*
Rep.	*Republic*
Soph.	*Sophist*

Theaet.   *Theaetetus*
Tim.   *Timaeus*

WORKS BY OTHERS

SCG   Thomas Aquinas, *Summa Contra Gentiles*
ST   Thomas Aquinas, *Summa Theologica*
In Ar. Top. Alexander of Aphrodisias, Commentary on
               Aristotle, *Topics*

Introduction

Aristotle regularly identifies philosophical concepts as *homonymous*. Although the force of this appraisal is not always immediately clear, it is surely noteworthy that when offering this sort of claim Aristotle very often has a critical target in sight. Most often, though not always, he has Plato in mind, because he believes that Plato systematically under-appreciates the complexity of core philosophical concepts. As a typical example, Aristotle argues that Plato incorrectly posits a single Form of Goodness which is the same for all good things; on this approach, every-thing which is good is good because it participates in a single, common Form, Goodness. According to Aristotle, Plato wrongly assumes that all good things are good in the same way, that their goodness is somehow common across all cases. When criticizing Plato for failing to appreciate homonymy, then, Aristotle means to assail this assumption by showing, as he often says, that goodness is 'spoken of in many ways' (*EN* 1096ª23–4). If Aristotle's criticisms are justified, Plato goes wrong in analysing Goodness, as in other cases, by assuming a form of unity unsus-tained by reflective philosophical analysis.

Although most prominent in critical contexts, homonymy finds equal employment in Aristotle's own positive philosophy. For he does not infer, as some later figures do, that complexity impedes philosophical analysis. On the contrary, he argues that certain homonymous concepts, while com-plex, are nevertheless ordered around a core. That is, Aristotle appeals to homonymy to find order in multiplicity; and he thinks that the order he finds permits genuine analysis of a sort which makes scientific inquiry legitimate and philosophical progress possible. Thus, for example, Aristotle evidently exploits the homonymy of being when arguing for the possibility of a unified science of metaphysics, whose object of study is being qua being rather than some portion of being considered in isolation (*Met.* 1003ª21–6). Aristotle's appeal to homonymy in this context is espe-cially significant, since he had earlier called into question the possibility of a science of being on the grounds that there is no genus of being (*An. Post.* 92ᵇ14, *Top.* 121ª16–19, ᵇ7–9). Aristotle evidently thinks that a science of being qua being is made possible at least in part by the recognition of being's homonymy (*Met.* 1003ª34–ᵇ10). This appeal may or may not be justified; in any event, it shows how Aristotle—at crucial junctures—supposes that homonymy has a clear and decisive role to play in his own positive philosophical theorizing.

In this book, I explicate and assess Aristotle's commitment to the homonymy of core philosophical concepts by evaluating his appeals to homonymy in both critical and constructive contexts. I argue that some of Aristotle's critical arguments are defensible, while others fall short of their intended marks. Similarly, on the constructive side, I maintain that some appeals to homonymy are philosophically fruitful, while others are uncompelling. In particular, I argue that Aristotle's commitment to the homonymy of being cannot be supported and that, consequently, one key feature of his mature ontology is problematic. Other concepts fare better. Most notably, I contend that Aristotle's account of the homonymy of goodness is defensible, although not precisely in the way he most prominently defends it. In any case, what is most significant, I argue, is that the methodology Aristotle develops and deploys in his constructive philosophy remains fruitful today. So while I question its application to being, I nevertheless agree that most of Aristotle's key appeals to homonymy can and should be defended.

FOCUS OF STUDY

In some ways, questions about homonymy in Aristotle are minutely focused and specialized; in other ways, they are highly general. They are specialized in so far as they require consideration of Aristotle's conception of dialectic, especially eristic dialectic; his approach to science and definition; his approach to the relation between concepts and properties; and especially his approach to *signification*. For, crucially, he thinks that difference in signification establishes homonymy. One central question, then, concerns Aristotle's attitude towards signification. If he thinks that signification is a broadly semantic relation, then he will also believe that difference in meaning establishes homonymy. If, by contrast, he introduces signification as a technical, non-semantic relation, involving essence specification in definition, then he will not be bound to suppose that meaning has anything at all to do with homonymy. In these ways, an adequate treatment of homonymy requires a full treatment of Aristotle's approach to signification, including, centrally, an investigation into the semantic status of significates in his system.

At the same time, the compass of these sometimes technical matters already suggests that questions about homonymy are also formidably general. To begin, the sweep of Aristotle's application of homonymy is in some ways astonishing. He appeals to homonymy in virtually every area of his philosophy. Along with being and goodness, Aristotle also accepts (or at times accepts) the homonymy or multivocity of: life, oneness, cause, source or principle, nature, necessity, substance, the body, friendship, part, whole, priority, posteriority, genus, species, the state, justice, and many others. Indeed, he dedicates an entire book of the *Metaphysics* to a recording and partial sorting of the many ways core philosophical notions are said to be. His preoccupation with homonymy influences his approach to almost every subject of inquiry he considers, and it clearly structures the philosophical methodology that he employs both when criticizing others and when advancing his own positive theories.

Viewed in this way, no comprehensive account of Aristotle's uses of homonymy could be complete without a case-by-case examination of his appeals. I do not undertake such a comprehensive account in this study. Instead, I try to examine the framework Aristotle develops for adjudicating disputes about homonymy by setting out its general features and by characterizing the metaphysical and semantic commitments necessary to establish the homonymy of a given philosophical concept in a disputed case. I consider in detail some of Aristotle's principal applications of homonymy, in a way intended to reveal both its variety and its force.

I approach this study with the belief that homonymy is a sort of lens through which we must view Aristotle's philosophy and in terms of which we must judge its ultimate success. At any rate, I seek to draw attention to the prominence Aristotle affords homonymy throughout his philosophical career. Even though, in the end, I will argue that we should not be persuaded by all of Aristotle's appeals to homonymy, I also contend, and hope to show by a consideration of cases, that contemporary philosophers ignore the apparatus of homonymy only to their own detriment.

PLAN OF STUDY

Aristotle's treatments of the homonymy of core philosophical concepts, including especially being and goodness, are sometimes highly abstract, and they must be understood as arising from the polemical contexts which motivate them. For these reasons, I consider these topics only after recounting Aristotle's general framework for introducing homonymy. Accordingly, I divide the study into two parts.

In Part I, I consider homonymy as such, mainly by reflecting on the uncontroversial cases upon which Aristotle himself relies when trying to explicate and motivate homonymy. I begin, in Chapter 1, by recounting Aristotle's introduction of homonymy in the *Categories*, settling some exegetical difficulties concerning his general conception of its nature. One such difficulty concerns the fact that Aristotle implicitly relies on different conceptions of homonymy in different contexts. Accordingly, I articulate and categorize the varieties of homonymy in his philosophy. In Chapter 2, I review Aristotle's methods for establishing homonymy, especially for those cases in which it is not readily apparent. I also challenge Aristotle's reliance on homonymy by motivating the sorts of criticisms an unsympathetic critic would be likely to mount. In general, these criticisms call for a method to be used in adjudicating disputed cases; for without such a method, disputed cases will quickly degenerate into stalemate. I focus on one such method in particular: Aristotle suggests that difference in signification is sufficient for homonymy. I do not presume that appeals to difference in signification will in every case settle disputes about homonymy, for in all but the simplest cases, disputes can equally arise concerning difference in signification.

Indeed, because signification is a technical notion for Aristotle, I turn to a discussion of signification itself in Chapter 3. I explore the connection between homonymy and signification, first by characterizing signification, and then by showing how appeals to signification undergird claims to homonymy. Part I closes with a general account of Aristotle's most striking commitment for the purposes of constructive philosophy: core-dependent homonymy. In Chapter 4, I sketch Aristotle's conception of core-dependence and address prominent problems with his approach.

In Part II, I investigate homonymy at work. I do not move through Aristotle's appeals to homonymy *seriatim*. Rather, I consider a very few cases, selected for their importance, interest, and representative character. In two cases, I urge that some of Aristotle's critics have failed to appreciate the power of homonymy in meeting objections to substantive Aristotelian theories. Thus, in Chapter 5, I show how Aristotle's appeals to the homonymy of the body permit him to meet otherwise forceful, foundational criticisms of his hylomorphic analysis of soul-body relations. Similarly, some of Aristotle's critics have understood him as accepting an ontology which countenances suspect entities, what I will call hyper-finely-individuated objects. (Very roughly, these interpreters understand Aristotle as accepting a commitment to the distinctness of Socrates, Socrates seated, Socrates cross-legged, the pale Socrates, the married Socrates, and so on for any ascription of a property to Socrates.) I argue in Chapter 6 that Aristotle does not accept such entities, because the homonymy of oneness permits him to say, as he should, that Socrates and Socrates seated are and are not one and the same.

Chapter 7 tests homonymy by developing a problem first broached in Chapter 1, though now in considerably greater detail. Aristotle claims several times that life is homonymous and that, consequently, it is impossible to provide a univocal account which covers all cases. Not all of Aristotle's interpreters regard these claims as his final thoughts on the matter; some think he eventually forges just such a univocal account. I argue, by contrast, that Aristotle does and should maintain that life is a core-dependent homonym. Indeed, his account of the homonymy of life provides a clear and compelling illustration of the fruitfulness of the methodology of homonymy.

The next two chapters take up Aristotle's most problematic appeals to homonymy, goodness, and being. I argue that Aristotle is correct to argue for the homonymy of goodness. Not all of his arguments for this conclusion are successful, however. Still, as long as one argument succeeds, Aristotle's claim will prove true. One noteworthy argument which fails evidently attempts to derive the homonymy of goodness from the homonymy of being. This argument fails because, as I argue in the final chapter, Aristotle cannot show that being is homonymous.

Aristotle's interpreters exhibit a striking disarray in their treatments of the homonymy of being. I consider some attempts to establish the homonymy of being, only to find them wanting. I then turn more minutely to Aristotle's own arguments, which proceed largely by appeal to technical features of his taxonomical methodology. These arguments also fail. Consequently, Aristotle fails to establish the homonymy of being.

Although I maintain that Aristotle cannot establish the homonymy of being, I do not infer that his commitment to homonymy as such is misguided. On the contrary, I maintain that outside this one application, Aristotle's commitment to homonymy is altogether well motivated; in particular, the method of definition it introduces is of genuine and lasting importance. At the very minimum, I argue, Aristotle is right to advocate homonymy as a form of constructive philosophical analysis. He has identified a framework which has too often been overlooked by those disenchanted with the prospects for genuine philosophical progress. Accordingly, I end Part II with a concluding afterword in which I appraise in a fully general way homonymy's enduring value.

I

Homonymy as Such

I

The Varieties of Homonymy

Aristotle opens an early work, the *Categories*,[1] by distinguishing between synonymy, homonymy, and paronymy (1ª1–15). I begin by considering this early characterization, which has come to be regarded as his canonical statement of the natures of homonymy and synonymy, with an eye toward assessing their eventual roles in his mature critical and non-critical philosophy. When first characterizing these notions, I avoid relying, as far as possible, on their most important and heavily disputed applications. I especially set aside Aristotle's treatments of being and goodness, the

---

[1] Some of Aristotle's interpreters, including e.g. Cherniss (1935) and Gill (1989), are *unitarians*, adopting a characteristically medieval hermeneutical method, according to which the extant corpus of Aristotle's works can be interpreted as adhering to a single, overarching philosophical system. On this approach, it is entirely appropriate to appeal indifferently to any given work by Aristotle to explicate remarks contained in another. More characteristic of the nineteenth and twentieth centuries is *developmentalism*, pioneered by Jaeger (1948) and upheld in various ways for portions of Aristotle's corpus by Solmsen (1929), Nuyens (1948), Gauthier and Jolif (1970), Owen (1965*b*), Wieland (1970), and Ross (1924 and 1957); developmentalists hold that Aristotle's views change and evolve throughout his productive life. In the current study I accept, without argument, a form of developmentalism, though not of the sort expounded by the developmentalists cited; the debate they have framed centres far too narrowly and constrictively on Aristotle's relationship to Plato. A better form of developmentalism recognizes that Aristotle accepts some features of Plato's metaphysics and epistemology while rejecting others throughout his life and focuses instead on the doctrinal and methodological dynamics internal to the progression of Aristotle's own thought. Graham (1987) appropriately relies on the introduction of hylomorphism as central to a shift in Aristotle's philosophical doctrine.

I accept the *Categories, De Interpretatione, Topics, Sophistici Elenchi, Eudemian Ethics, Prior Analytics*, and *Posterior Analytics* as comparatively early. I accept the *Physics, Nicomachean Ethics, De Anima, Politics*, and most of the *Metaphysics*, as comparatively late. Strikingly, regarding the topic of this study, it seems clear that Aristotle developed an account of homonymy very early and that he relied upon some notion of homonymy— whether or not this notion itself varies—from his earliest to his latest works. In a general way, it is safe to say that Aristotle found appeals to homonymy appropriate in nearly every subject he investigated in virtually every period of his life. More difficult is the question of whether his appeals to homonymy always come to the same; and more difficult still is the question of whether Aristotle's occasional reversals about whether a given concept qualifies as homonymous reflect a change in attitude about the concepts in question or a broader shift in his attitude about homonymy itself. I approach these questions as they arise in the text.

discussions of which are theoretically loaded in ways that make them unsuitable as sources of relatively neutral data concerning homonymy and multivocity.[2] Instead, I begin by concentrating on the kinds of examples Aristotle himself employs when illustrating these notions. My principal aims are: (i) to unearth his precise concept of homonymy; (ii) to expose the theoretical foundations Aristotle supplies for its applications; and (iii) to reflect on the cogency of the doctrine thus displayed.

As introduced in *Categories* 1, the distinction between homonymy and synonymy is crucially unclear: most notably, Aristotle does not indicate in this passage whether homonymy and synonymy are exhaustive. Accordingly, I turn to additional texts in order to establish how Aristotle understands the distinction. I argue that Aristotle relies on importantly distinct forms of homonymy for different purposes and that, consequently, he intends homonymy and synonymy to be understood exhaustively. That is, since he deploys different forms of homonymy in different contexts, he implicitly recognizes forms which he could not if he also recognized some *tertium quid* between homonymy and synonymy. Every definable term is, therefore, either homonymous or synonymous.[3]

---

[2] 'Multivocity' renders Aristotle's expression 'spoken of in many ways' (*pollachôs legomenon*). I follow Aristotle's dominant practice in using homonymy and multivocity interchangeably, as in the whole of *Top.* 1. 15, as well as at *Top.* 129^{b}31, 148^{a}23; *An. Pr.* 32^{a}18-21 with 25^{a}37–^{b}2; *Phys.* 186^{a}25–^{b}12. It may seem that he sometimes treats homonymy and multivocality as distinct; I argue below in 1. 3 that despite this appearance, homonymy and multivocity are indeed co-extensive.

[3] It will be noticed that in the following, I make little use of Aristotle's appeals to analogy. I do not think Aristotle regards either homonymy or analogy as species of the other. In this I agree with Aubenque (1978), although his concern is mostly restricted to the analogy of being. I am also entirely in agreement with the last sentence of an insightful passage from Wiggins (1971, 32 n. b): 'One of Aristotle's favourite clear cases of focal meaning, *healthy*, happens to provide a particularly good example of failure of unitary analysis. It is exactly similar to the failure with *good*. The only way of representing *healthy* as unitary is to write the lexicon entry as e.g. *pertaining to the health or well-being of an organism*. This collects up *healthy complexion, healthy constitution, healthy weather, healthy climate, healthy drink* all right but only at the cost of allowing such oddities as *healthy hospital, healthy cur, healthy lecture, healthy textbook*. *Pertaining to* is too vague. But to specify something more exact involves splitting the lexicon entry for the word. The true account of the matter is that the various senses of healthy are related by being arranged in different ways around the "focus" health. It is no accident that all the things which are properly called healthy are so called but the ways are different ways. And this explanation has nothing to do with analogy.' I provide a non-accidental account of relations in 4. 4, and it is one which makes no use of analogy. In some passages Aristotle may appear to offer analogy as a kind of homonymy, e.g. *EN* 1096^{b}27–8. In these passages, however, Aristotle is better understood as contrasting analogy with homonymy, in particular what I call associated homonymy. Thus, at *EN* 1096^{b}27–8, Aristotle wishes to know whether goodness should be regarded as an associated homonym, or *rather* according to analogy: *all' ara ge tô(i) aph' henos einai ê pros hen hapanta suntelein, ê mallon kat' analogian.*

1.1 ARISTOTLE'S INTRODUCTION OF HOMONYMY

When Aristotle first introduces the related notions of homonymy and synonymy in the *Categories*, he claims: 'Those things are called homonymous of which the name alone is common, but the account of being corresponding to the name is different . . . Those things are called synonymous of which the name is common, and the account of being corresponding to the name is the same' (1^a1–4, 6–7). Aristotle's account of synonymy is relatively straightforward:[4] 'x and y are synonymously F iff (i) both are F and (ii) the definitions corresponding to "F" in "x is F" and "y is F" are the same.' But his initial account of homonymy is ambiguous between *discrete* and *comprehensive homonymy*:

> Discrete Homonymy (DH): x and y are homonymously F iff (i) they have their name in common, but (ii) their definitions have nothing in common and so do not overlap in any way.

> Comprehensive Homonymy (CH): x and y are homonymously F iff (i) they have their name in common, (ii) their definitions do not completely overlap.[5]

According to CH, homonymy and synonymy are exhaustive; according to DH, they are not. DH holds that the names of homonymous things are what we would call ambiguous, so that river banks and savings banks are homonymously called 'banks'. Definitions of 'bank' in these occurrences have no overlap. By contrast, if we think that the definitions of 'healthy' as it occurs in 'healthy people', 'healthy complexions', and 'healthy food' ought somehow to be related, then if we adopt DH, there must be some *tertium quid* between homonymy and synonymy to capture this connection. CH admits these as cases of homonymy.

DH and CH agree about the domain of homonymy. They equally hold

---

[4] The general framework for the approach to homonymy employed in this section is indebted to the accounts offered by Irwin (1981) and Fine (1988); my analysis differs in significant regards from the accounts offered in those articles.

[5] The qualification that 'they do not completely overlap' is necessary to ensure that the distinction between synonymy and homonymy remains exclusive, as Aristotle evidently thinks it is. (The matter is not completely clear however; see Williams 1982, 113.) Importantly, in saying that these distinctions are exclusive, I do not thereby suggest that two homonymous things cannot be synonymous as well. Rather, two things can be both synonymous and homonymous only in so far as they fall under sortals which are not co-extensive. Thus, decoys used in duck hunting and the ducks themselves may be homonymously ducks, but synonymously targets or synonymously magnitudes. See Ackrill 1963, 71.

that homonyms are things and not words.[6] As stated, both reflect Aristotle's requirement that homonyms have the same name, but not the same account of being (*onoma monon koinon, ho de kata tounoma logos tês ousias heteros*; *Cat.* 1ª1–2). According to the formula of the *Categories*, then, homonyms are homonyms just because they share names but not accounts or definitions. This is almost precisely the formula Aristotle applies elsewhere as well. In the *De Anima*, for example, he holds that eyes incapable of functioning as eyes are homonymous eyes because they lack the being of an eye with respect to its account (*ousia ophthalmou hê kata ton logon*; *An.* 412ᵇ19–20). Here again, no mention is made of ambiguity or multiplicity in sense.

If two things are homonymous only if they have the same name, then trivially two words are homonymous only if they have the same names. It is possible for a word to have a name; but it would be rare and unusual. It would be even rarer and more unusual for two words to have the same name. Hence, if we rely exclusively on Aristotle's introduction of homonymy in the *Categories*, no words will turn out to be homonyms, even though as a technical matter it would be possible for two words to satisfy either DH or CH.

Some caution is required here, however. In agreeing about the domain of homonymy, DH and CH seem to contradict Aristotle's contention elsewhere that words can be homonymous.[7] For example, in a discussion of

---

[6] Ackrill (1963, 71) observes: 'The terms "homonymous" and "synonymous", as defined by Aristotle in this chapter [*Cat.* 1], apply not to words but to things.'

[7] There is, consequently, a controversy over whether Aristotle intends homonymy to mark different senses of words or advances it rather as a doctrine about properties. The dominant interpretation, of Owen (1960), Leszl (1970), and Hamlyn (1977–8), treats homonymy as a semantic phenomenon and as pertaining to word meaning. The alternative of Ackrill (1963, 71 *inter alia*) and Irwin (1981) holds that Aristotle's primary concern is with things, not words. Cf. also Charlton (1970, 54). Irwin argues that homonymy is indicated by difference of definition, where Aristotelian definitions signify essences, while: 'an essence is not a meaning; it is a real property, a real feature of the world, an Aristotelian universal. A homonymous name can be replaced by many definitions because it signifies many real properties, not because it has different meanings' (535). Irwin argues as follows: if Aristotle's appeal to homonymy were merely an appeal to difference in sense, he would be content to appeal 'to what is accessible to the normal competent speaker of the language' (543); but Aristotle 'is not primarily concerned with the ordinary competent speaker's judgments of different senses, but with how things really are' (543). Although Irwin is surely right about the methodological principle concerning common usage, it is not clear that he is right to infer from that that Aristotle is not therefore interested in senses of words. Given the approach to signification which I think Aristotle prefers (see Chapter 3), there will be no hard distinction between senses of words and essences or real features of the world: meanings will be, for natural kinds, essences. Competent speakers of the language will in some cases not have grasped these meanings; but then they will not have fully grasped the senses of certain words (e.g. 'piety'). Given their false beliefs about the things to which the words they use apply, they will therefore not be in a position to adjudicate questions of

the nature of acting and being acted upon in *De Generatione et Corruptione*, Aristotle notes that acting and being affected are not possible in the strict sense without contact. Consequently, any clear account of these matters must first consider the nature of contact. As he turns his attention to the nature of contact, Aristotle, in his characteristic way, encourages sensitivity to its homonymous aspect: 'So one must first speak about contact. Just as nearly all other words are used in many ways, some homonymously, and others on the basis of other, prior [words], so it is with "contact"' (*GC* 322^{b}29–32).[8] This passage is noteworthy in several respects. First, Aristotle clearly and indisputably refers to words (*onomata*) as homonymous.[9] This will seem surprising if we understand either CH or DH to contain Arisotle's canonical definition of homonymy; on the other hand, it will seem unsurprising if we have been supposing that homonymy is akin to ambiguity and so a feature of words. In either case, Aristotle's claim here is perfectly general.[10] He says without qualification that 'contact' is like many other words, indeed like 'nearly all other words', in having a variety of senses. This provides evidence that homonymy pertains to senses of words and so suggests an alignment of homonymy with ambiguity. Consequently, we are not entitled to settle on either DH or CH as Aristotle's official view at this stage, at least not as regards the domain of homonymy. Aristotle may wish to accept the position he enunciates in the *De Generatione et Corruptione* and so proceed to state DH

homonymy, even though they will still be appropriately regarded as competent speakers of the language. I will argue that the dominant view is right to suppose that Aristotle has a concern about the senses of words, but wrong if it takes this to entail that Aristotle is *merely* concerned about how words are used or *merely* interested in common usage; Irwin is right to criticize this tendency in the dominant view but not therefore licensed to infer that homonymy is not 'intended to mark different senses of words' (524). Thus, I prefer an account which recognizes what is right in both the dominant and alternative approaches. I take up this issue more fully below in 2. 4, in 2. 5, and especially in 3. 5. In these early chapters, I will follow Arisotle's own practice of sometimes calling words homonymous and sometimes calling non-linguistic entities homonymous.

[8] On this use of homonymy see Williams 1982, 113.

[9] The *ta men . . . ta de* at 322^{b}31 clearly pick up *onomata* as their antecedent.

[10] The periphrastic translation of Joachim (1922b, 141) is incorrect: 'Now every term which possesses a variety of meanings includes those various meanings *either* owing to a mere coincidence of language, *or* owing to a real order of derivation in different things to which it is applied. This may be taken to hold of Contact as of all such terms' (italics as found). Joachim seems to ignore the *hekaston* at 322^{b}30, thus giving the impression that Aristotle's remark is restricted to a class of multivocals of unspecified size which divide into homonyms or words with related senses. The Revised Oxford Translation (Barnes 1984) overcompensates: 'Now no doubt, just as every other name is used in many senses (in some cases homonymously, in others one use being derived from other and prior uses), so too is it with contact.' Williams (1982, 21–2) best relates Aristotle's claim: 'Just as almost all other words are used in a variety of senses, some of them equivocally, others with some of their senses dependent on other, prior senses, so it is with "contact".'

and CH more carefully. Alternately, investigation of DH and CH in the context of Aristotle's semantic theory may yet point to a way of resolving the apparent tension.

1. 2 DISCRETE AND COMPREHENSIVE HOMONYMY

However we are to resolve questions concerning the domain of homonymy, it is possible to determine independently whether Aristotle favours DH or CH. Because the canonical formulations of homonymy and synonymy in *Categories* 1 are compatible with both CH and DH, it will be useful to consider Aristotle's illustrations. Unfortunately, even the examples of homonymy Aristotle provides following his introduction of the terms do not establish conclusively whether he intends DH or CH: 'For example, both a man and a picture <of an animal> are animals; only their name is common, while the account of being with respect to the name are different. For if someone were to specify what being an animal is for each of them, one would specify a distinct account for each' (*Cat.* 1ᵃ2–6). The examples may seem initially to support DH: an account of what it is for a man to be an animal should have nothing in common with an account of what it is for a picture of an animal to be an animal, since a man is an animal and a picture of an animal not an animal at all but a picture. That is, someone might easily understand Aristotle as suggesting that the accounts of 'animal' as applied differently to humans and pictures will be distinct. The account of 'animal' as applied to humans will make reference to the possession of a perceptual soul which Aristotle thinks is definitive of all animals. By contrast, pictures lack souls altogether and so cannot be so defined in any such terms.

This would be a mistake. Presumably for something to qualify as a picture of an animal, as opposed to e.g. a picture of a jar of nails, it will have to represent, in some sense, something with a perceptual soul.[11] So, far from supporting DH, Aristotle's example here evidently lends support to

---

[11] Still, it is worth noting that there are two distinct predicates in question: (i) 'being a picture'; and (ii) 'being a picture of a man'. If we intend to offer an account of the predicate in (i), there is no reason to suppose any connection, because the predicate 'being a picture' will have nothing in common with the predicate 'being a man'. If we want to offer an account of the predicate in (ii), (a) it will need to be an account in one of the two looser senses identified in *Met.* vii. 4 (1030ᵃ17–23, ᵇ4–12), and (b) the reason we will have association will derive from the fact that we have a case of representation. It is primarily in cases where we have *resemblance* or *imitation* where we may be inclined to find association where it does not exist.

a kind of homonymy which recognizes definitional overlap and so actually reflects a commitment to CH.[12]

Looking beyond *Categories* 1, it becomes clear that Aristotle sometimes introduces instances of homonymy which would be consistent with his maintaining only DH, while at other times he clearly requires a more comprehensive notion. Since CH is broader than DH (DH is a special case of CH, and as such is recognized as a special case of CH in clause (ii) of the definition provided), Aristotle must reject DH as a general account of homonymy in favour of CH.

The scope of Aristotle's conception of homonymy emerges very clearly in *Topics* i. 15, the chapter in which Aristotle introduces his standard homonymy indicators.[13] In this chapter Aristotle provides an overview of techniques for determining whether something is spoken of in many ways or in only one (106ᵃ9). Some of the tests recommended are explicitly linguistic:

First one should see if the opposite is spoken of in many ways, whether the divergence is in form or in word (*ean te tô(i) eidei ean te tô(i) onomati diaphônê*). For in some cases, the differences <emerge> immediately even in the words, e.g. in the case of voice the opposite of sharp is flat, while in the case of mass the opposite of sharp is blunt. It is clear then that the opposite of sharp is spoken of in many ways. But if this is so, then sharp too is spoken of in many ways. For corresponding to each of those, the opposite will be different, since the same sharp will not be the opposite of flat and blunt, even while sharp is the opposite of each . . . Similarly, the opposite of fine in the case of an animal is ugly, but in the case of a dwelling the opposite of fine is wretched, so that fine is homonymous. (*Top.* i. 15, 106ᵃ10–22, with omission)

In some cases, however, linguistic tests are not immediately decisive (106ᵃ23–35). We say that the opposite of a bright light is a dim light; but

---

[12] Unfortunately, this passage does not conclusively support CH, since it is perhaps possible that Aristotle is remarking on the fact that *zô(i)on* applies to animals and to pictures of any sort, that is that *zô(i)on* might mean 'animal' or 'picture'. In that case, Aristotle will be pointing out that one word might have definitions which do not overlap in any way. If so, then his illustration would be consistent with his maintaining DH. Ackrill (1963, 71) alludes to this point. See also Owen 1965a, 73–5.

[13] Leszl disputes the value of this and other crucial passages from the *Topics*, 'which anyway should not be taken too seriously, since they belong to an early work' (1970, 335). It is not possible, however, to develop any sharp criterion for distinguishing Aristotle's account of homonymy in the earlier and later works. Certainly the *Categories* is early (on dating, see n. 1 above); there the fully technical notion of homonymy is already introduced; the *Eudemian Ethics*, also very early, already exploits the apparatus of Aristotle's most complex doctrines of homonymy. It is difficult to see how either of these treatments of homonymy could be dismissed for belonging to an early work. In my view, the same holds true for the *Topics*. It is, contrary to Leszl's suggestion, striking how constant Aristotle's framework for homonymy remains throughout his earlier and later works.

we equally say that the opposite of a bright pupil is a dim pupil. Even so, a bright light is luminous, and a bright student intelligent. The linguistic test of opposites does not reveal homonymy in this case: 'In some cases there is no divergence in names, but the difference in form is immediately quite obvious' (106^a23–5).[14]

Aristotle provides an assortment of homonyms in *Topics* i. 15: fine, sharp, clear, obscure, pleasure, love, seeing, not seeing, perceiving, just, justly, health, healthy, healthily, good, donkey, balanced, and colour.[15] The sheer variety of cases makes it difficult to see what single account underlies the concept. In some cases, Aristotle seems to require only DH, since the instances of homonyms are really accidents of language. Most notably, donkeys are homonymous, since some donkeys are animals and some are engines (*Top.* 107^a18–23). (Aristotle trades on the fact that *onos* (donkey) is used both of the animal and of a type of pulley; a close English analogue would be *crane*.) In this case, Aristotle applies the formula of *Categories* 1, noting that 'the accounts corresponding to the name are different' (107^a20). Indeed, he uses this example as one where items in different genera share a name. In the first case, the item is in the genus animal; the other is not an animal, but a machine. Since no two things can have the same account without falling under the same genus, the cranes in question must be homonymous. And since these cranes have accounts which do not overlap in any way, DH captures the type of homonymy they illustrate.

Many other examples in the chapter require a notion which does not segregate the accounts of homonyms so tidily. Two examples are particu-

---

[14] On the tests for homonymy, see 2. 4–2. 5. In the present context, it is worth recalling that many of these are linguistic tests, including: (i) one use of a term has an opposite, but another lacks an opposite altogether (106^a36–b3); (ii) a set of opposed terms has an intermediary taken one way, but none at all taken another (106^b4–9); (iii) a set of opposed terms has one intermediary taken one way, and another a host of intermediaries taken anther (106^b9–12); (iv) if the polar opposite of a term has more than one sense, that term will itself have more than one sense (106^b13–20); (v) in cases where terms are opposed as privations and possessions, homonymy in the privation will result in homonymy in the possession and *vice versa* (106^b21–8); (vi) homonymy in one inflection (*ptôsis*) of a term results in homonymy in other inflections of that term (106^b29–107^a2).

[15] Some of these examples need to be adapted in order to construct tests in English parallel to those Aristotle employs. It is worth emphasizing that the lack of a cross-language fit already indicates the linguistic character of Aristotle's tests for homonymy and multivocity. Because the tests provide indications of homonymy without constituting the phenomenon, however, it would be premature to conclude on the basis of them alone that homonymy pertains exclusively to meaning. Even so, the tests show that meaning will have something to do with homonymy; for if meaning is a guide, Aristotle must suppose there is a non-accidental relation between word meaning and homonymy. Otherwise the tests would be altogether otiose.

larly noteworthy. First, Aristotle offers 'seeing' as homonymous and as illustrating a procedure for detecting homonyms by determining whether the contradictory of a term is homonymous. He claims that since 'not seeing' (*to mê blepein*) is used in several ways, the same must hold true of 'seeing' (*to blepein*). One might fail to see for either of two reasons: (i) because one lacks the power to see or (ii) because one fails to activate one's power to see. Thus, the sentence

(1) She doesn't see the traffic.

might mean she cannot see the traffic, because she is blind, or simply that she does not see it at the moment, because she is distracted or somehow hindered. We might not notice that 'She sees' is similarly ambiguous, since when used transitively 'see' does not normally display ambiguity. When uttered in normal circumstances, the negations of (1) ('She sees the traffic') would not be taken to mean that she possesses the ability to see the traffic. Still, the sentence 'She sees' can be used in more than one sense, corresponding to the senses of the negation. Hence, Aristotle concludes, because 'not seeing' is used in several ways, 'it is necessary that *seeing* is also used in several ways' (*Top.* 106ᵇ17–18).

Now, focusing on the homonym uncovered, we see that Aristotle cannot employ only DH. For surely any account of actually seeing we provide will be closely connected to our account of possessing the power of sight. Indeed, an account of actually seeing now will make reference to the capacity of sight (*Top.* 106ᵇ18–20). Conversely, Aristotle believes that any account of the capacity of sight must be given in terms of actual sight, which in turn is given partially in terms of the objects of sight. The distinction between two types of actually seeing is familiar from the *De Anima* (417ᵃ9–14); and the principle that an account of a capacity will make reference to the actuality of that capacity follows from Aristotle's commitment to a general principle according to which potential *F*s are defined in terms actual *F*s (*An.* 415ᵃ18–20; *Met.* 1049ᵃ10–ᵇ2, 1049ᵇ12–17, 1071ᵇ12–1072ᵃ18). Hence, the accounts of seeing as a capacity and as an actualization of that capacity appeal to one another.

In consequence, 'seeing' is recognized as a homonym. According to DH, homonyms have accounts which do not overlap in any way. Therefore, 'seeing' is not homonymous in any way countenanced by DH. Since he recognizes 'seeing' as homonymous, and since he accepts either DH or CH, Aristotle must therefore embrace CH as his general account of homonymy.

A second example from *Topics* i. 15 equally establishes this point. As a further means for detecting homonyms, Aristotle recommends determining whether opposites are related as privation and disposition (*sterêsis kai hexis*), in which case if one of the opposites is homonymous, the other is too (106ᵇ21–8). For example, if 'perceiving' (*to aisthanesthai*) is homonymous as applied to body and soul, being 'imperceptive' (*to anaisthê-ton*) will also be homonymous. This illustration is a bit more obscure than some others in the chapter, but presumably Aristotle means that just as we speak of perception as a cognitive act, so too can we employ 'imperceptive' as a psychological epithet or as a literal description of a body's powers. In any case, perceiving is homonymous as applied to the body and soul. If both applications are literal, then the accounts must be very closely related indeed, perhaps even the material and formal specification of the same act. If perception is applied slightly more metaphorically ('He's not terribly perceptive'), even then the accounts will be centrally connected. Like 'seeing', 'perceiving' is a homonym not recognized by DH. Therefore, Aristotle's admission of it as a case of homonymy provides further evidence that he accepts CH and not DH. Indeed, that Aristotle suggests it as a paradigmatic case of homonymy capable of illustrating a procedure for uncovering homonymy when it may not be apparent suggests that it is for him a perfectly uncontroversial instance of the phenomenon. Here again we see Aristotle's implicit rejection of DH and his consequent acceptance of CH.

Aristotle's attitude toward homonymy as such provides further confirmation. Both within *Topics* i. 15 and elsewhere Aristotle counsels that we must be careful not to overlook cases of homonymy. Because some cranes fall into the genus of animal and other cranes are machines, no one could overlook the homonymy of 'crane'. This point is really quite general: no one could miss the vast majority of the homonyms DH countenances, precisely because these homonyms have accounts with no overlap. Such homonyms have names as starkly ambiguous as 'bank'. Consequently, if Aristotle did not recognize homonyms beyond those DH admits, he should not worry terribly about anyone's ever overlooking cases of homonymy.

Yet he does worry that we are sometimes oblivious to homonyms. Here there are two sorts of cautions. In the first case, we may satisfy ourselves that certain things are not homonymous, because, having analysed them, we see no divergence in account. Aristotle counsels against our resting content at the first level of analysis:

Homonymy often trails into the accounts themselves unnoticed, and for this reason one needs to look into the accounts. For example should someone say that what is indicative or productive of health is what has balance with respect to health, one should not stand fast but must inquire further in what way balance was mentioned in each case, for example, if in the one case it means to be of such a sort as to produce health, but in the other it means to be of such a sort as to be able to indicate what kind of state <health> is. (*Top.* 107ᵇ6–12)

Homonyms entering only at the second tier of analysis render the original terms homonymous. Thus, 'health' as it occurs in 'indicative of health' and 'productive of health' is evidently homonymous,[16] since balance is the homonymous portion of the account 'having balance with respect to health'. Here homonymy 'trails in unnoticed' (*lanthanei parakolouthoun to homônumon*), because we think we have a univocal account, which upon closer examination turns out to contain a homonymous term. Aristotle's caution here reveals an attitude towards homonymy he should not have if he accepts DH. Hence, again, Aristotle must accept not DH but CH as his general doctrine.

Aristotle's second type of caution is more direct. Although he notes that in many cases 'the difference in form is immediately quite obvious' (*Top.* 106ᵃ23–5), Aristotle equally warns that we slip into fallacious reasoning because we miss subtle homonyms. In *Sophistici Elenchi* 6, for example: 'An error occurs with respect to homonymy and an account[17] by not being able to distinguish what is spoken of in many ways; for it is not easy to distinguish some cases, for example, one and being and same' (169ᵃ22–5). Similar remarks occur later, in the *Sophistici Elenchi* 33:

Just as in those <arguments that go awry> with respect to homonymy, which seem to be the simplest type of fallacy, some being clear to just anyone (for nearly all humorous remarks depend upon their articulation . . .),[18] while others escape the notice of even the most experienced (an indication of this is that even <the most experienced> often fight about words (*machontai pollakis peri tôn onomatôn*), whether, for example, 'being' and 'one' mean (*sêmainei*) the same in all cases or something different. For to some it seems that 'one' and 'being' mean the same, while others solve the argument of Zeno and Parmenides by saying that 'one' and 'being' are used in many ways); so it is similar concerning fallacies[19] of coincidence and each of the other sorts: some of the arguments will be easy to see, but

---

[16] Hintikka (1971, 368) raises the question of whether it is possible for whole phrases to count as homonyms, or whether in accordance with *Soph. El.* 7, 169ᵃ22–5 they should be regarded as cases of amphiboly. Probably Aristotle regards the whole phrase as homonymous, but only because one word in it is so. See n. 27 below.

[17] The Revised Oxford Translation (Barnes 1984), together with Ross in the OCT (Ross 1958), finds *logos* obscure and refers to an analogously obscure use at *Soph. El.* 168ᵃ25. Probably Aristotle has in mind the problem of homonymy creeping unnoticed into accounts (*logoi*), as discussed at *Top.* 107ᵇ6–12. His point here then may be paraphrased: 'Error occurs with respect to homonymy even in an account because of our not being able to distinguish what is spoken of in many ways within the account itself.'

[18] There follows a series of Greek jokes which do not carry over into English very well.

[19] Reading *tôn* with Ross (1958) at 182ᵇ28, as against the manuscripts.

others more difficult, and to grasp what sort of fallacy it is, and whether it is a refutation or not, is not equally easy in all instances. (182ᵇ12–31, with omission)

As opposed to the obvious homonyms that make for jokes and the like, other homonyms escape the notice of even those most experienced at ferretting them out. This would be altogether improbable if Aristotle's conception of homonymy were restricted to DH.

A final instance, perhaps even more compelling, comes from *Nicomachean Ethics* i. 6. In discussing the homonymy of goodness, Aristotle identifies by name the class of homonyms regarded as obvious and as rife in jokes in his treatments in the *Topics* and *Sophistici Elenchi*. He claims that *good* 'at any rate does not resemble those <homonyms which are> homonyms by chance' (*ou gar eoike tois ge apo tuchês homônumois*, 1096ᵇ26–7). Such homonyms are homonyms by chance (*apo tuchês*) in the sense recognized by DH: their accounts do not overlap in any way, even though they have the same names. It is, therefore, an accident of language that in these cases items referred to as *F*s are all called *F*. Consequently, we may infer in a neutral way that Aristotle recognizes non-chance homonyms. We may conclude, further, that he seems to recognize them at every stage in which he recognizes homonyms at all. Since it makes no provision for non-chance homonyms, DH cannot be Aristotle's general account of homonymy. Therefore, CH best captures Aristotle's general conception of homonymy.

All of these instances reveal that Aristotle recognizes forms of homonymy which DH fails to countenance. They therefore show that Aristotle accepts CH. Further reflection on the purpose of Aristotle's introduction of the apparatus of homonymy and synonymy into the *Categories* helps to cement this conclusion.

Readers of the *Categories* should be initially perplexed by both the abruptness and the disjointedness of Aristotle's opening chapter. He introduces homonymy and synonymy, but he fails to say why; and he never mentions homonymy by name again anywhere in the work, though, significantly, he does employ the notion of multivocity (*Cat.* 8ᵇ26). He mentions synonymy only once, when pointing out that entities falling under the same substance sortal and differentia are spoken of synonymously (3ª33–ᵇ10).[20] (Thus, horses are called horses synonymously; horses and cats are called animals synonymously.) Why, then, does he open the work by introducing a distinction between homonymy and synonymy?

Given the anti-Platonic purport of the categories, Aristotle very likely

---

[20] Synonymy is not mentioned in the companion *De Interpretatione*, and homonymy is mentioned only twice, once non-substantively (17ª35) and once substantively (23ª7).

introduces homonymy because he wishes to encourage sensitivity about both inter- and intra-categorial non-univocity. He thinks that every item in every one of the categories is a being; but he does not believe that being is used univocally of them. Instead, he recognizes four types of beings, corresponding to the types of predications: some things are said-of and in other things; others are said-of but not in; others are in but not said-of; and still others are neither said-of nor in (*Cat.* 10^{a}16–^{b}9). Answering to these four types of predication conditions are four types of entities: non-substance universals; substantial universals; non-substance particulars; and individual substances, the primary substances (*Cat.* 10^{a}16–^{b}9; cf. 2^{a}11–19). Each of these kinds is irreducibly distinct from the others, since, by a simple appeal to Leibniz's law, it can be shown that no one type can be identified with any other. Aristotle further believes that he can infer from this irreducible difference a distinction in being itself, such that what it is for a primary substance to be will be distinct from what it is for e.g. a non-substance universal to be. That is, the accounts of *is* in:

(1) Speusippus is.

and

(2) The colour blue is.

are distinct. Thus, the categories provide a premise in an argument for the homonymy of being. The introduction of homonymy, then, conditions the reader for this form of inter-categorial non-univocity.

Without reflecting on the efficacy of any general argument for the homonymy of being at this stage,[21] it can be appreciated that Aristotle thinks that the types of beings are connected around primary substances, without which none of the other types of beings would exist. Since the accounts of *is* in (1) and (2) are distinct but nevertheless connected, Aristotle's introduction of homonymy in the first chapter of the *Categories* already assumes that a form of association is possible; consequently, DH does not reflect Aristotle's full posture regarding homonymy. Therefore, Aristotle accepts CH.

A similar conclusion follows regarding intra-categorial homonymy. Aristotle recognizes primary and secondary instances within several categories, including substance and quantity (*Cat.* 5^{a}38–^{b}1). If the substantial universal, man, and Speusippus both count as substances, but only the latter counts as a substance in the primary way, then accounts of *substance* in:

---

[21] I explore this argument below in Chapter 9.

(3) Speusippus is a substance.

and

(4) Man is a substance.

will differ. This, if correct, is already sufficient to establish that *substance* is homonymous, without discriminating further between the various types of homonymy.

Still, since an account of 'substance' as applied to the species will make reference back to primary substances, it follows that applications of 'substance' in (3) and (4) are associated. Here too, then, in the case of intra-categorial homonymy, we see Aristotle implicitly relying on CH. Even though 'substance' and 'quantity' are spoken of in several ways, all ways lead back to primary instances.

Aristotle's reliance on these forms of homonymy both explains his introduction of homonymy in *Categories* 1 and reveals his early accept-ance of CH. Aristotle introduces a distinction between homonymy and synonymy in order to alert readers to two instances of non-univocity, first, between items within individual categories and, second, between beings across all the categories. Because both intra- and inter-categorial instances of homonymy are clearly associated, Aristotle's introduction of homonymy and synonymy into the *Categories* supports our earlier argu-ments that Aristotle endorses CH rather than DH. They add to those arguments a further finding: Aristotle accepts CH from his earliest writings.

1.3 NON-HOMONYMOUS MULTIVOCALS

If these arguments are correct, then the distinction between synonymy and homonymy ought to be exhaustive. This conclusion is threatened by some of Aristotle's practices: sometimes he contrasts homonyms with multivocals, either implicitly (*Top.* 110^{b}16–32) or expressly (*Met.* 1003^{a}33–4, 1030^{a}32–^{b}5; *GC* 322^{b}29–32). Yet we expect multivocals to be non-synonymous. If there are multivocals which are non-homonymous, but also non-synonymous, then the distinction between homonymy and synonymy is not exhaustive. If this distinction fails to be exhaustive, Aristotle will not be able to establish homonymy on the basis of mere non-univocity. For then non-univocity could be sufficient for non-synonymy without also being sufficient for homonymy. Thus, multivocity could fail to establish homonymy; and so far we have been following Aristotle's general practice of treating homonyms and multivocals as extensionally equivalent.[22]

Different kinds of cases present themselves. When determining whether some domain admits of a single science, Aristotle sometimes denies homonymy while awarding multivocity (*Met.* 1003^{a}33–4); when he does so, he clearly at the same time denies synonymy. But he also sometimes identifies a class of multivocals which are not homonymous (*Top.* 110^{b}16–32); in these cases, though, there may be a temptation to understand him as recognizing a class of synonymous multivocals. If the first appearance is genuine, then the contrast between synonymy and homonymy is not exhaustive. If the second appearance is genuine, then non-univocity will not always establish homonymy. In this case, we will have to distinguish between those appeals to multivocity which suffice for homonymy and those which are consistent with synonymy. Both cases, in different ways, challenge the contention that the homonymy/synonymy distinction is exhaustive.

The worry about Aristotle's appeals to multivocity can also be put this way: homonymy and synonymy are mutually exclusive; we can establish homonymy from non-synonymy only if they are exhaustive; we can establish non-synonymy from multivocity only if multivocity is the same as or is entailed by non-univocity. If he recognizes some synonymous multivocals, Aristotle denies this last entailment.

The most noteworthy instance of the first sort of case occurs when Aristotle inaugurates the science of being qua being by insisting that 'being is spoken of in many ways, but with respect to one thing and some one nature and not homonymously' (*Met.* 1003^{a}33–4). Here he claims flatly that being is a multivocal, but not homonymous. Hence, being may seem to be either a synonymous multivocal or neither synonymous nor homonymous.

It is clear that this inference is mistaken. For, first, being is not

---

[22] Matthews (1995, 235) doubts that this is Aristotle's actual practice: 'It seems plausible to say that, according to Aristotle, any word used homonymously is also said in more than one way. But we are not required to make the reverse inference. We need not suppose, that is, that Aristotle supposes any term said in many ways is therefore used homonymously. We are free to suppose that "said in many ways" is a looser classification—one that includes, but is not restricted to, cases of genuine homonymy.' Matthews does not comment on Aristotle's practice of very often simply using the terms interchangeably—where the interchange, as in *Topics* i. 15, esp. 106^{a}1–8, heads in both directions indifferently. I therefore agree with Owen (1960, 182 n. 5): 'If a *word* is *pollachôs legomenon* then it is a case of homonymy, requiring different definitions in different uses.' Owen goes on to draw an exception for *phrases*, which he holds can be *pollachôs legomena* without also being homonymous. I disagree, but the disagreement is not relevant to the present considerations.

synonymous. But, second, it is not therefore neither homonymous nor synonymous.[23] Instead, as we have seen, it is possible for being to be one sort of homonym without its being another sort. That is, since Aristotle maintains CH and not DH, he can easily contrast multivocals with some homonyms without denying that all multivocals are themselves homonyms. Further, given that the examples of the sort of multivocity Aristotle has in mind for being are paradigmatic associated homonyms, namely health and medicine (*Met.* 1003^{a}34–^{b}2; cf. 1030^{a}35–^{b}3, 1061^{a}10–11), it follows that he intends here only a contrast between being and non-associated homonyms, that is, discrete homonyms. Since this contrast is compatible with treating being as comprehensively homonymous, Aristotle nowhere denies that multivocals are homonyms. On the contrary, he introduces stock examples of associated homonymy as illustrative of the form of multivocity he has in mind.

Taking all this together: cases like being do not threaten the exhaustiveness of Aristotle's distinction between homonymy and synonymy.[24] Nor do they introduce a class of non-homonymous multivocals. For they are homonymous.

Instances of the second sort are not so easily handled, for they may appear to be synonymous multivocals. If this appearance is correct, then even though Aristotle's contrast between homonymy and synonymy will be exhaustive, multivocity and homonymy will no longer be co-extensive. In this case, any direct inference from multivocity to homonymy will be insecure. For it will also have to be shown that the multivocals in question are non-synonymous.

Two cases require consideration. Aristotle treats the possible (*endechesthai*) as homonymous (*An. Pr.* 32^{a}16–21; cf. *Int.* 22^{b}29–23^{a}20 for related remarks about potentialities). It is possible for some things to be *F* without its also being possible that they be not-*F* (this is one-sided possibility, e.g. it is possible for a fire to be hot, but not possible for fire not to be hot), while it is possible for some other things to be *F* or not-*F* (this is two-sided possibility, e.g. it is possible for Socrates to be ill or not to be ill). So far, then, multivocity and homonymy march in step in the normal way.

Aristotle complicates matters by drawing a further distinction among

[23] Ross (1923, i. 256), surprisingly, misses this point, supposing that being must be paronymous, because it is 'intermediate between' homonymy and synonymy. This is evidently because he thinks the *Categories* recognizes only discrete homonyms. Since we have shown that Aristotle accepts CH and not DH, we are free not to follow Ross in this inference.

[24] The general remarks offered here hold equally for oneness and goodness (*Met.* 1045^{b}6, 1053^{b}22, 1059^{b}33; *EN* 1096^{b}27). Alexander (242.5) rightly adds other associated homonyms to the list.

forms of possibility. Beyond one- and two-sided possibility, he wants to distinguish possibilities which attach to subjects, for the most part, from those which happen only by chance:

These distinctions having been made, let us say in turn that the *possible* is spoken of in two ways (*to endechesthai kata duo legetai tropous*), [i] what occurs for the most part and falls short of necessity, e.g. that a man greys, or grows or declines, or generally what belongs naturally <to a subject> (for this is not a consequence: that it has necessity because of man's not existing forever, even though *if* a man exists, it is either of necessity or for the most part <that he declines>), [ii] the other is indefinite, which is able to be both thus and not thus, e.g. that an animal walks or an earthquake occurs while it is walking, or generally things occurring by chance; for it is not naturally more this way than the opposite. (*An. Pr.* 32ᵇ4–13)

In the first case, a man's hair turns grey not of necessity, but for the most part. So, it is possible for a man's hair to turn grey, though this is not a mere one-sided possibility. Likewise, an animal may or may not walk, and if it walks, it may or may not do so during the occurrence of an earthquake. Its walking during the occurrence of an earthquake is, of course, also a two-sided possibility; but it is not one explicable by laws, as a man's turning grey is (*An. Pr.* 32ᵃ18–23). The occurrence is freakish. As Aristotle suggests, there is no general inclination or disposition for an animal to walk when an earthquake occurs, and no general tendency for an earthquake to occur while an animal walks.

However that may be, it is striking in the present context that Aristotle introduces this second distinction by noting that the 'possible' is spoken of in two ways (*kata duo legetai tropous*), where this evidently qualifies it as a multivocal. But since both ways of speaking eventually appeal to two-sided possibility, this may appear to be an instance of a synonymous multivocal.[25] For since both ways of speaking appeal to the same account, we have a case of synonymy.

It is unlikely that Aristotle has in mind here an instance of synonymous multivocality. For it is unlikely that he intends the two forms of 'possibility' to be synonymous. Certainly, at any rate, this does not follow merely from the fact that both involve one form of possibility (two-sided); this would be necessary but not sufficient for synonymy. Rather, he sees 'possible' as distinct in:

(5) It is possible that Portnoy turns grey as he ages.

and

---

[25] Irwin (1981, 530) offers this sort of approach.

(6) It is possible that the milkman delivers milk as leaves whirl in the yard.

The distinction, he thinks, arises because we have a basis for regarding the possibility in (5) as law-governed, or law-regulated, whereas the possibility in (6) is mere happenstance. We have reason to predict (5), but none to predict (6). Proposition (5) reports a natural possibility, while (6) reports an indefinite possibility (*An. Pr.* 32^{b}14–16).

Because (5), but not (6), is law-regulated, the form of possibility involved is robust possibility. The form of possibility invoked in (6) is, by contrast, slender possibility. Though robust and slender possibilities agree in being two-sided, they nevertheless are distinguishable in terms of their frequency, predictability, and explicability. This is why Aristotle thinks that robust possibilities have a role to play in syllogistic, whereas slender possibilities have none (*An. Pr.* 32^{b}18–22). This is also why slender and robust possibilities are usefully conflated for rhetorical effect. ('Now you admit that it is possible that the brakes may have malfunctioned, though earlier you denied it!') The discernible senses, though equally two-sided, are distinct. Accordingly, the accounts answering to them are distinct, however related. 'Possible' is, therefore, not synonymous, but homonymous. Consequently, Aristotle does not recognize an instance of synonymous multivocity in 'possible'.

Much the same conclusion can be drawn with respect to the second case of seeming synonymous multivocality. In *Topics* ii. 3, Aristotle recommends vigilance about 'however many things are spoken of in many ways, not according to homonymy, but in some other way' (*hosa mê kath' homônumian legetai pollachôs alla kat' allon tropon*, 110^{b}16–17). Perhaps we should infer that if the 'other way' is not synonymy, then the contrast is not exhaustive.[26] Aristotle's example is difficult. He offers 'one sciene of many things' (110^{b}17) as illustrative. Evidently, this phrase is spoken of in many ways because some one of its constituents, either 'one' or 'many', is spoken of in many ways.[27]

---

[26] Irwin (1981, 529–30) seems to reason this way. At any rate, he concludes: '[Some non-homonymous multivocals] are not intermediate cases between synonymy and homonymy, but [are] synonymous; Aristotle can still maintain that all non-synonymous multivocals are homonymous.'

[27] We should probably agree with Irwin (1981) as against Owen (1965a) and Hintikka (1973) that, properly, Aristotle has just one word in mind. Still, we need not agree with him that it is inappropriate to treat the entire phrase as spoken of in many ways; it may be spoken of in many ways in an inherited way. Indeed, for compositional reasons, it may be possible for a phrase to be spoken of in many ways even if none of its words is; this will be true when the phrase is not amphibolous (*Rhet.* 1407^{a}39, *Soph. El.* 166^{a}7–14).

In the face of this example, some have been tempted to recognize a class of synonymous multivocals,[28] while others may suspect an example of a *tertium quid* between homonymy and synonymy. Neither of these options is attractive.

Fortunately, neither option is necessary. First, and for familiar reasons, it is possible that Aristotle means only that the phrase is not discretely homonymous. This is nevertheless compatible with its being homonymous, because it is compatible with its being what I have been calling, informally, an *associated homonym*, that is, a homonym whose constituents are related in identifiable ways. Hence, we need not regard this phrase as a sort of *tertium quid* between homonymy and synonymy. Even so, someone might accept this judgement without agreeing that the phrase is homonymous. For someone might argue that the phrase is a multivocal even though it is synonymous.[29]

The argument would be as follows. Suppose the whole phrase is multivocal because one of its words, perhaps 'one', is multivocal.[30] Aristotle argues that 'one science is of many things' is multivocal on the grounds that a given object of study may consist in an end or a means, or in two distinct ends, or in an object essentially or accidentally construed (*Top.* 110ᵇ16–25). He evidently has a number of different cases in mind. Considering the last one first, a triangle may be an object of a science investigating either equilaterals or figures whose interior angles equal two right angles. These sciences must be distinct, since they have disjoint domains. Even so, one science may consider objects common to both. This is one science of many things because its objects cannot be co-extensive (*Top.* 110ᵇ26–31). At the same time, as the first cases (concerning ends) makes clear, it is possible for a science to be one science of many things in other ways. Medicine, for example, is one science of many things because it studies both the end, the production of health, and the means to that end, including diet (*Top.* 110ᵇ16–17). Since these objects are distinct, medicine is one science of many things. This way of being one science of many things is not the same as being one science of things which are only accidentally the same. Hence, 'being one science of many things' is multivocal.

Now, however, it is maintained that neither any portion of the whole phrase, nor the whole phrase itself, is homonymous. Certainly, 'many'

---

[28] Irwin (1981, 530).

[29] Irwin (1981, 530) argues this way.

[30] Irwin thinks that 'many' is likely the source of the multivocity. I suspect, on the basis of *Top.* 110ᵇ33–7, that Aristotle has 'one' in mind. Perhaps, though, these alternatives should not be regarded as competitors. Indeed, one might argue that it is the interaction between them which produces multivocity.

retains one account across various applications; so too does 'one'. Treating the whole phrase as multivocal, then, we find no obvious homonymy between

(7) Medicine is one science of many things.

and

(8) A branch of geometry is one science of many things.

Even so, it is possible for Aristotle to have in mind not synonymous multivocals, but rather multivocals which are associated homonyms. For this is the sort of multivocality, he tells us, which easily goes unnoticed (*Top.* 110^{b}34); and it is associated, not discrete homonymy, which often escapes us. Hence, as he encourages, a more careful analysis of (7) and (8) may reveal just this form of homonymy.

Perhaps Aristotle thinks that the whole phrase is homonymous, either because some one of its words is, or because, when interacting in various applications, they engender homonymy. He may imagine someone arguing that: (i) medicine is one science over many things; (ii) so too is a given branch of geometry; hence, (iii) these sciences have something in common. In arguing against such a point of view, Aristotle will want not to concede synonymy, for then his opponent will have a point. Instead, he will want to point toward unnoticed homonymy, according to which the accounts of the predicates of (7) and (8) will be distinct, though related.

If this is correct, then we do not have a clear instance of a non-homonymous multivocal. For these phrases, being spoken of in many ways is neither some *tertium quid* nor a form of synonymy. It is, rather, an instance of associated homonymy.

All these cases of purported non-homonymous multivocals turn out to be non-problematic. They do not force any *tertium quid* between homonymy and synonymy. Instead they confirm CH by turning out to be cases of homonymy. This result is important in another way as well, since it further supports our initial inclination to follow Aristotle's standard practice of treating homonyms and multivocals as co-extensive. Indeed, the examples considered lend support to the stronger judgement that they are necessarily co-extensive, since all multivocals turn out to be, non-accidentally, discrete or associated homonyms. We will, consequently, be entitled to infer that if *F*s are spoken of in many ways, those same *F*s are homonymous; we will further be justified in treating Aristotle's occasional contrasts between homonymy and multivocity as contrasts between multivocity and discrete homonymy only. There are no non-homonymous multivocals.

1.4 DISCRETE, NON-ACCIDENTAL HOMONYMS

Summing up thus far: CH divides homonyms into two types: (i) those whose accounts do not overlap at all, and (ii) those whose accounts overlap to some degree but not completely. We have implicitly already introduced the second type of homonyms as *associated* homonyms. We may call all homonyms of the first type *discrete* homonyms, since they correspond exactly to all the homonyms there would be if DH captured all the homonyms recognized by Aristotle. In the next two sections, I explore some of the features of discrete homonymy. I then turn to some features of associated homonymy.

Aristotle describes discrete homonyms as obvious, or simple, or silly.[31] Indeed, he sometimes gives the impression that all discrete homonyms are easily apprehended as such and that only associated homonyms hold any philosophical interest. This is unfortunate, since Aristotle is elsewhere keen to point out that some discrete homonyms are nevertheless non-accidentally homonymous. Moreover, in just these cases the homonymy involves a characteristic often reserved for associated homonyms: their homonymy is difficult to detect.

Two clear examples come from the *De Anima*, one concerning an artefact and another a part of the body. First, after offering a general account of hylomorphism, Aristotle compares an ensouled body with a functioning axe:

It has been stated in a general way what the soul is. For it is a substance in accordance with the account. For this is the essence of such a body. Just as if another body, say an axe, were an organic body, what it is to be an axe would be its substance, and this would be its soul; for if this were separated, it would no longer be an axe, except homonymously. But as it is, it is an axe <and not an organic body>. (*An.* 412ᵇ10–15)

If an axe lost its axe-ity, and so was no longer able to do what it is that axes do, it would no longer be an axe. There are several ways an axe might lose its ability to function as an axe, some of them external to its structure (if it were locked in a vault no one could open) and some of them internal (if the handle decomposed to the point where it could no longer be used to hoist the axe for chopping). If an axe's internal constitution decays to the point where it can no longer do what axes do, then Aristotle thinks it is no longer an axe, 'except homonymously' (*plên homonumôs*). An

---

[31] Aristotle says that such a homonym is *euethes*, where the precise pejorative character is difficult to specify. Probably he means that they are mistaken only by the dim-witted.

account of an axe that is an axe only homonymously will make no appeal
to what it is to be an axe (*to pelekei einai*), since to be an axe is to be a certain sort of artefact capable of chopping.[32]

It will nevertheless be perfectly natural to speak of such an artefact as
an axe. Indeed, if we had any particular reason to pick out a given axe by
a name or name-like singular term, we would naturally use the same name
during the periods when it is an axe and not an axe except homonymously.
'Grandpa's axe' might equally designate the axe once used to chop wood
but now kept around as a memento of its former owner. This axe is not
called 'axe' by a mere accident of language, in the way that various banks
are called 'banks'. Even so, on the assumption that it is no longer functional, its account will differ from the account of a functioning axe. Hence,
although it is a discrete homonym, its homonymy will not be silly or obvious; nor will it be felicitously assimilated to the chance homonyms (*apo
tuchês*) identified at *Nicomachean Ethics* 1096ᵇ26–7, if this is to suggest
that it is a matter of mere linguistic happenstance that it has the name it
has.

Aristotle makes a similar observation slightly later in the *De Anima*
when applying general soul-body hylomorphism to the hylomorphism of
particular sense organs:

It is also necessary to contemplate <the ramifications of> what has been said for
the parts of the body. For if the eye were an animal, sight would be its soul, for
this is the being of the eye with respect to its account (*ousia ophthalmou hê kata
ton logon*). The eye is the matter of sight; if sight is lacking, it is no longer an eye,
except homonymously, just as a stone eye or a painted eye (*An.* 412ᵇ18–22).

An eye which cannot see is not an eye, except homonymously. Here
Aristotle draws out the point that an eye incapable of seeing does not have
the account of an eye by placing non-seeing human eyes into the same category as the eyes in statues and paintings. We would be tempted to refer
to a sightless person's eyes as 'eyes', and indeed, this reference would be
wholly justified in common discourse. If someone loses his sight, then we
may well ask, 'Is there perhaps any hope of an operation on your eyes to
enable you to see again?' In assimilating sightless eyes to representations

---

[32] It is natural to suppose that an account of an axe which cannot chop will make reference to an account of an 'axe', that is that an axe and an ex-axe will be related. Here
Aristotle seems to disagree, by relying on the thought that the essences of things are *functionally specified*, so that an ex-axe which cannot cut—which does not fulfil the function of
axes—will not qualify as an axe at all. This, of course, flouts some linguistic conventions.
This example raises a question about what forms of connection are necessary and sufficient
for associated homonymy. I explore some of these questions in more detail when considering a form of association central to Aristotle's thoughts about homonymy in 4. 3–5 below.

of eyes, Aristotle knowingly cuts against the grain of common usage; this is the point of using uncontroversial examples of homonymous eyes as homonyms of the same class of eyes whose homonymy we might otherwise easily overlook.

The examples mentioned are not exceptions for Aristotle. He regularly places the body and bodily parts into the second category by aligning them with sculptures (*An.* 412^{b}17–22, *Meteor.* 389^{b}25–390^{a}13, *Pol.* 1253^{a}20–5, *PA* 640^{b}30–641^{a}6, *GA* 734^{b}25–7). In these contexts, he often uses the locution *x* is not an *F*, or *x* is no longer an *F*, 'except homonymously' (*plên homônumôs*; e.g. *An.* 412^{b}21); he sometimes makes the same point by saying that something is not an *F* 'or <is> rather <an *F*> homonymously', *all' ê homônumôs* (e.g. *An.* 412^{b}14–15). In so speaking, he means that the *F*s in question have nothing definitionally in common with genuine *F*s, and are called *F*s only by custom or courtesy. These are discrete homonyms which nevertheless form a class worthy of our attention; unlike the puns mentioned at *Sophistici Elenchi* 33, they will not be 'clear to just anyone'.

Aristotle will require a special reason for insisting that we recognize these forms of homonymy. Since they will rarely make any practical difference (in the normal course of events, no one will run into difficulty when calling a non-functioning axe an axe), Aristotle needs to argue that philosophical mistakes may result from a failure to recognize this sort of homonymy. He will also need to supply an argument that such a form of homonymy even exists.

Aristotle offers such an argument by attempting to derive this form of homonymy from a deeply held thesis of kind individuation, which involves what I will call *functional determination*.

1.5 DISCRETE, NON-ACCIDENTAL HOMONYMY AND FUNCTIONAL DETERMINATION

Discrete, non-accidental homonyms of philosophical significance invariably involve *F*s which are at one time genuine but at another time spurious. They cease being genuine *F*s while maintaining the outward form or appearance of genuine *F*s.[33] In such cases, although the accounts are

---

[33] By 'outward form or appearance' I mean roughly what Aristotle would in some contexts call the *schêma* as opposed to the *morphê* or *logos*. See e.g. *Phys.* 193^{b}6–12 where they are implicitly contrasted. An especially relevant passage in this regard is *PA* 641^{a}18 ff., where Aristotle points out that a soulless animal is not really an animal except with regard to shape (*schêma*). More generally, at *PA* 640^{b}29 ff., he criticizes Democritus for holding that the form of a thing is to be identified with its external figure. Thus, I am sympathetic

discrete, it would be a mistake to say that it is a mere accident of language that the *F*s in question are called *F*s. Because of entrenched patterns of use, we continue to call something an *F*, perhaps without even realizing that it is no longer a genuine *F*. This may lead us astray in philosophical contexts, and so should be avoided. Perhaps the errors we will make will not be as severe as the errors we will make when confusing associated homonyms; but they will be errors none the less.[34] And they will have their roots in failures to distinguish genuine *F*s from *F*s which are non-accidental, discrete homonyms.

Aristotle's methodological caution in this regard presupposes a metaphysical distinction between genuine and homonymous *F*s. In saying that their accounts differ, Aristotle commits himself to an analysis of accounts which yields a real distinction among the kinds of things we call *F*s. He provides such an analysis by offering a thesis of functional determination for kind membership and individuation, a thesis whose ramifications for homonymy Aristotle fully appreciates. In *Politics* i. 2, when illustrating the priority of the polis to the family and individual, Aristotle alludes to his view about the homonymy of non-functioning body parts, and he illuminates it further by situating it in a more general account of kind individuation:

Further, the polis is by nature clearly prior to the family and to the individual, since the whole is of necessity prior to the part. If the whole body is destroyed, there will be no foot or hand, except homonymously, as we might speak of a stone hand; for when destroyed, the hand will be no better than that. But things are defined by their function and power, and we ought not to say that they are defined the same when they no longer have their proper quality, but only that they are homonymous. (*Pol.* 1253ᵃ19–25)

Bracketing Aristotle's political purposes in appealing to the homonymy of bodily parts,[35] we see that homonymous *F*s become such when they lose the function or power we associate with being an *F*. Indeed, Aristotle says that *all* things are defined by their functions, so that something will be

with Hartman (1976, 554 n. 15): 'Probably even in the case of a statue the shape is not a necessary and sufficient condition for its being the sort of thing it is, whether a statue of someone in particular or just a statue.' I would strengthen Hartman's claim by adding that the shape of a statue is not necessary *or* sufficient for something's being the particular statue it is, or even to its being a statue at all. In the one case, something could certainly be a statue of Lenin without having his precise shape, and arguably without having even his approximate shape; in the other, neither Lenin nor Lenin's corpse is a statue of Lenin.

[34] See Chapter 5 for a discussion of errors resulting from failures to notice discrete, non-accidental homonyms.

[35] See Miller (1995, 44–50) for a discussion of the role of functional determination in Aristotle's political theory.

defined as an *F* only when it has the function and power of a genuine *F*.

This same claim is put still more perspicuously in *Meteorologica* iv, where the connection to homonymy is once again stressed:

All things are defined by their function: for <in those cases where> things are able to perform their function, each thing truly is <*F*>, e.g. an eye, when it can see. But when something cannot <perform that function> it is homonymously <*F*>, like a dead eye or one made of stone, just as a wooden saw is no more a saw than one in a picture. The same, then, <holds true> of flesh. (*Meteor.* 390ª10–15; cf. *GA* 734ᵇ24–31)³⁶

Here Aristotle straightforwardly asserts a thesis of *functinal determination* (FD):

> FD: An individual *x* will belong to a kind or class *F* iff: *x* can perform the function of that kind or class.

FD makes it both necessary and sufficient for membership in kind *F* that *x* have the function definitive of being *F*. This reflects Aristotle's contentions that 'if something can perform its function, it truly is <an *F*>' (the sufficiency claim) and that 'when something cannot <perform that function> it is homonymously <*F*>' (the necessity claim).

FD may strike us as just right for a full range of artefacts. What makes something a light is its ability to illuminate. Fluorescent and incandescent bulbs are equally lights, as are gas lanterns, oil lamps, campfires, candles, and automobile headlights. These entities have different structures and are made of a wide range of different stuffs. Still, we are justified in calling them all *F*s because of their shared function. In these cases, the sufficiency condition binds together an otherwise motley group into a discernible class. The necessity condition seems equally apt in these cases. If we are in a dark room with an empty oil lamp, we are in a room without a light.

FD may nevertheless seem too sweeping, and some scientific essentialists will surely reject it when thinking of natural kinds. If we identify water with H_2O, then we will not believe that something with all of water's

---

³⁶ Several points must be made about this passage from the *Meteorologica*: (1) it comes from a work whose authenticity has been doubted; (2) some translators regard it as locally qualified (it occurs in a passage concerning homoeomerous bodies); (3) it states in an unrestricted form what elsewhere Aristotle tacitly restricts to classes with clear functions (*erga*), e.g. artefacts. But these observations do not undermine its utility here: (1) the *Meteorologica* is not spurious, or at any rate Book iv is genuine (cf. Gaiser 1985, 53–7); (2) the language Aristotle uses suggests that he intends a general rather than qualified principle, and he elsewhere in similar contexts extends it beyond this restricted class, e.g. *PA* 640ᵇ18–23 (and even within this passage it is unclear why Aristotle would mention the eye if he were concerned only with homoeomerous bodies); and (3) human beings and mental states will certainly be included in any restricted formulation of the claim that function determines kind membership. Cf. *Met.* 1029ᵇ23–1030ª17 and *EN* 1098ª7–8.

macroscopic properties is water, even if we think that the stuff in question can, to a deep level of analysis, perform all the functions we normally associate with water.[37] Here we think of water as a compositional stuff rather than as a functional stuff. Perhaps food is a functional stuff exclusively. (At any rate, if we go to twin earth, we will have a hard time denying that our twins are eating food, even if we discover that its sub-molecular structure is quite unlike the sub-molecular structure of our food.) Aristotle's essentialism does not easily accommodate essentialism built around compositional stuffs.[38] Even so, since it is offered as a central thesis about kind membership and individuation, FD requires us to disassociate *F*s whose accounts differ, even when these differences matter only in philosophical contexts and come to little for non-philosophical, pragmatic reasons.

Aristotle believes that drawing distinctions among discrete homonyms will sometimes be necessary to fend off objections to a proposal which seems otherwise defensible. For example, we do not need in ordinary contexts to think of 'body' as homonymous. Indeed, it may seem perverse of Aristotle to insist that just as a non-functioning eye is an eye only homonymously, so a corpse is not really a human body at all. This after all contravenes established patterns of discourse, in the sense that it prohibits one from saying e.g. 'Lenin's body lies in Red Square'. Aristotle's recognition of the body as a discrete, non-accidental homonym puts him at odds with customary linguistic practice. We call *S*'s corpse her body; yet Aristotle insists that the corpse is not her body, except homonymously. Does Aristotle flout customary use idly?

This is unlikely, given his own tendency to rebuke those who are willing to ignore customary linguistic patterns without sufficient motivation (*Top.* 110ᵃ14–22). Instead, Aristotle's willingness to depart from customary linguistic use in technical philosophical contexts reflects his awareness of the ramifications of his views on homonymy for non-technical contexts.[39]

We see, then, that Aristotle's recognition of discrete, non-accidental homonyms serves him in philosophical contexts where we might otherwise think he has erred. In these contexts, FD provides a principle for dis-

---

[37] It is necessary in these contexts to distinguish—as Aristotle fails to distinguish—between functional kinds and functional specifications of kinds, which may or may not be functional kinds. Functional kinds are kinds whose essences are exhausted by their functions; it is possible that a compositional kind, whose essence is not so exhausted, be functionally specified as, e.g., alkaline may be specified as *the substance which turns the litmus paper blue*. We should not therefore treat alkaline as a functional kind. See Ludwig 1994.

[38] For an opposed view, see Bolton 1976, together with the criticisms of Ackrill 1981.

[39] In Chapter 6 I explore in detail one such context.

tinguishing discrete homonyms whose utility becomes apparent only in contexts where homonymy may go unnoticed, despite its being discrete. Linguistic practice disinclines us to recognize these homonyms, and for most pragmatic purposes linguistic practice is justified in this regard. Still, philosophical purposes may demand greater precision than pragmatic purposes and may therefore warrant our recognizing a functionally determined class of discrete, non-accidental homonyms.

1. 6 ASSOCIATED HOMONYMY

Of greater philosophical interest are *associated* homonyms. These are homonyms, recognized by CH (ii), whose accounts overlap, but not completely.[40] We have seen that although most discrete homonyms are patently homonymous, there may nevertheless be some non-accidental instances which are so subtle that they elude all but those who have a special interest in noticing them. Their homonymy may emerge only in rarefied contexts. Associated homonyms will go unnoticed more frequently still. Their homonymy may escape even those most alert to the possibility of unnoticed equivocity.

The reasons why discrete and associated homonyms tend to go unnoticed are, however, mainly distinct. We have seen that non-accidental discrete homonyms require a special principle of kind individuation, the functional determination thesis, to establish their non-univocity. In these cases, the accounts should not make defining reference to one another. In the case of associated homonyms, we may expect some similar principle of differentiation. Aristotle does not, however, supply one. Instead, he uses some of the linguistic tests of *Topics* i. 15 to show non-univocity, even when common sense may overlook or even deny it. In these cases, common sense may overlook or deny homonymy for a perfectly good reason: the notions distinguished may be so close that it is difficult to appreciate

---

[40] It is possible for someone (i) to accept the defintion of homonymy offered in the *Categories*, (ii) to recognize the possibility of associated homonymy, but to insist (iii) that associated homonyms need not associate around some core. Thus, there is logical space for non-core dependent associated homonyms (perhaps these will be akin to the sort of homonymy Walker (1979) encourages, without success, for 'friendship'; see Chapter 2, n. 28). Because Aristotle never explores or exploits this possibility in any detail, I will mainly set it aside. For clarity's sake, though, one can divide homonyms exhaustively into those which are associated and those which are discrete, allowing that there may be species of each.

that they are indeed distinct. In this sense, their very association renders associated homonyms hard to grasp.

A good example is Aristotle's treatment of justice in *Nicomachean Ethics* v. 1. In this discussion, Aristotle realizes that not everyone will appreciate that justice is non-univocal. Even so, he insists that we would be wrong to insist on univocity:

> Justice and injustice seem to be spoken of in many ways, although this escapes our notice because of the extreme closeness of their homonymy. These cases are unlike cases where <the homonymy> is far apart (for here the difference in form is great), as for example <it is clear> that the collar bones of animals and that with which we open doors are homonymously called 'keys'. (*EN* 1129ª26–31)

We call the flesh around the teeth and the sticky substances secreted by some trees 'gums'; but we never think that pine tar holds teeth in place. In these cases the homonymy is far apart. By contrast, Aristotle thinks, we call two different virtues—one concerned with the common good and the other concerned with avoiding rapacious greed (*pleonexia*)—'justice' (*EN* 1129ª14–ᵇ19). We would be wrong to treat these as one and the same virtue; but we would also be wrong to treat them as discrete in the way that gums are discrete. These are cases of multivocals which are nevertheless associated.

Aristotle rightly sees that he must argue for the non-univocity of justice. His argument employs a test of contrariety: 'for the most part it follows that if one <pair of> contraries is spoken of in several ways so is the other, for example, <if> what is just <is spoken of in several ways, so> too is what is unjust' (*EN* 1129ª23–5). Employing this principle, Aristotle argues that since we say that *S* can be unjust in more than one way, when we say that *S* is just, we may be making one of a number of different sorts of judgements:

> Let us, then, determine in how many ways a person is spoken of as unjust. Both the lawless person and the person who is rapacious and unfair seem to be unjust, with the clear result that both the lawful and the fair person will be just. Consequently, what is lawful and what is fair will both be just, while what is lawless and what is unfair will be unjust. (*EN* 1129ª31–ᵇ1)

By applying the test of contraries, Aristotle thinks he can show that 'justice' is spoken of in several ways.

His argument is direct: (i) if *x* can be not-*F* in more than one way, then *x* can be *F* in more than one way as well; (ii) *x* can be not-just in more than one way; (iii) hence, *x* can be just in more than one way; (iv) hence, 'justice' is spoken of in more than one way. A person can be unjust by being lawless or by being unfair; hence, a person can be just by being lawful or by being fair.[41] Hence, if we merely judge that someone is just, we may fail to discriminate more precisely what it is that we mean to claim. If the first premise of Aristotle's argument is apt, this will be due to our failure to recognize multivocity.

However that may be, Aristotle does not intend for us to regard 'just' as applied twice over to the same person as somehow discrete. On the contrary, he suggests, without developed argument, that they are intimately connected (*EN* 1130ª33–5). Surely his intuitions are correct. When we say that Hannah is just because she is attentive to the common good, we are saying something closely related to what we say when we claim that she is just because she does not covet the fairly acquired goods of her friends. These are both states of character; they are both states concerned with the good; and they are both directly relevant to our appraisal of Hannah's virtue. The accounts of justice in these cases will involve significant, but not total, overlap. Here we have an instance of associated homonymy.

Aristotle does not, however, claim definitional priority for one form of justice over the other. In this way, his practice contrasts markedly with a more philosophically engaging kind of associated homonymy, what I will call *core-dependent* association. Aristotle's favourite examples of core-dependent homonymy illustrate a kind of association which is especially useful in offering accounts of central philosophical concepts. '[E]verything which is healthy is related to health (*pros hugieian*), some by preserving health, some by producing health, others by being indicative of health, and others by being receptive of health' (*Met.* 1003ª34–ᵇ1). In offering health as an example of core-dependent association, Aristotle thinks that some things should be obvious about health from a moment's reflection on the following statements:

(9) Socrates is healthy.

(10) Socrates' complexion is healthy.

(11) Socrates' regimen is healthy.

(12) Socrates' salary is healthy.

Considering first (9)–(11), two things seem evident: (i) the predicate 'is healthy' has different meanings in the different statements; and (ii) the

---

[41] In calling someone 'lawful' (*nomimos*) Aristotle evidently does not mean merely 'law-abiding'. His contrast is therefore not merely between someone who is conventionally just and someone who is always fair. Rather, he means to distinguish between someone who is mindful of the common good (*EN* 1129ᵇ15–19) and someone who wishes to avoid rapaciousness directed toward a particular person (*EN* 1130ª14–ᵇ7). Sophocles captures the difference between the two forms of justice in his presentation of Odysseus in the *Philoctetes*.

different meanings are nevertheless closely related. Adding (12) into consideration, we see first that the predicate has another distinct meaning, but we perhaps sense that this meaning is not tied as closely to the meaning of the predicate in (9) as are the meanings of the predicate in (10) and (11).

Aristotle's motivation for regarding some central philosophical concepts as homonymous mirrors the impulses we have for differentiation and co-ordination in (9)–(11). An account of the predicate 'is healthy' in (10) will advert to the state an organism is in when it is free of disease and functioning well. By contrast, accounts of the predicate in (10) and (11) will not advert directly to a state of the subjects of the predications, the complexion and the regimen, at all. They will appeal instead to the fact that good complexions tend to be produced by healthy bodies and so tend to indicate health in healthy bodies, or to the fact that regimens count as healthy because they tend to produce and preserve health in healthy bodies. Thus, roughly, substituting these provisional accounts into (9)–(11), we have:

> (9′) Socrates is in a state of being free of disease and functioning well.

> (10′) Socrates' complexion indicates that Socrates is healthy.

> (11′) Socrates' regimen brings it about that Socrates is healthy.

Because both (10′) and (11′) appeal to the notion of health itself, as realized in Socrates, we can avoid circularity by appealing to the account of health provided in (9′), yielding:

> (10″) Socrates' complexion indicates that Socrates is in a state of being free of disease and functioning well.

> (11″) Socrates' regimen brings it about that Socrates is in a state of being free of disease and functioning well.

The accounts of the predicates in our original (10) and (11) are therefore neither reducible to the account of the predicate in (9) nor independent of it. Rather, they must appeal to it in order to be correct and complete; but since they mean more than the predicate of (9) taken alone, such an appeal is insufficient by itself.

Aristotle thinks the methodology suggested by these sorts of examples has a very wide range of application in philosophical theorizing. Most notably, he holds that although philosophical accounts of being and goodness are forthcoming, such accounts will not be univocal, as some others have thought. On the contrary, their accounts will exhibit a genuine and ineliminable multiplicity. Still, if being and goodness are like health, they should nevertheless exhibit order in multiplicity: there must be a core to being and goodness to which accounts of non-core instances necessarily appeal.

These sorts of homonyms must be non-univocal and thus equivocal; but they will not be *merely equivocal*. Nor will they be merely associated, for they will be associated in a way which commands a special philosophical interest. Taking all this together, then, although equivocal, some homonyms will be related in ways which turn out to be philosophically significant. These homonyms will be associated homonyms. Among associated homonyms, some turn out to be especially important, because they display a core dependence which helps us discern the heart of the concept, despite its non-univocity.

I. 7 SEDUCTIVE AND NON-SEDUCTIVE HOMONYMS

Although most discrete homonyms are patently non-univocal, the non-accidental may be quite subtle. Their homonymy may emerge only in rarefied contexts. Associated homonyms share in common with discrete, non-accidental homonyms the ability to confound even those most sensitive to homonymy as such. We appreciate this point by reconsidering Aristotle's approach to justice in *Nicomachean Ethics* v. 1 and emphasizing more strongly a feature we have already touched upon. After allowing that 'justice' is spoken of in many ways, Aristotle allows that this often 'escapes our notice because of the extreme closeness of their homonymy' (*dia to suneggus einai tên homônumian autôn lanthanei*, 1129ª26–7). Homonyms are more or less obvious depending upon the degree of closeness.

Aristotle does not specify what constitutes closeness in homonymy. His examples initially suggest a distinction parallel to his earlier distinction between associated and discrete homonyms: keys are discretely homonymous, since their accounts have no overlap, whereas types of justice admit of considerable overlap. No one misses the homonymy of 'key'; many have overlooked the homonymy of 'justice'. If this is correct, we have a further distinction among homonyms: seductive versus non-seductive homonyms, which sorts homonyms along a distinction already drawn between discrete and associated homonyms.

This alignment, however, cannot be correct. Since we have already recognized a class of discrete homonyms whose homonymy is apt to escape our notice, seductive homonyms must cut across the distinction between discrete and associated homonyms. Hence, we have the following account of seductive homonymy:

> SH: x and y are seductive homonyms iff: (i) they have their names in common; (ii) their definitions (a) have nothing in common, or (b) do not overlap completely; and (iii) they so appear that they tend to induce observers to suppose that (ii a) is false, when it is true, or (ii b) is false, when it is true.

Aristotle provides keys as an example of a distant or non-seductive homonymy. These are non-seductive homonyms which are also discrete homonyms. Their accounts do not overlap in any way, and it would be unlikely for anyone to regard them as synonyms.

One could, of course, construct an abnormal circumstance in which they would be seductive for some person. This is because, as clause (iii) in the SH makes explicit, something's being a seductive homonym will be partly a function of the sophistication of a perceiver. Hence, seductive homonymy will be both subjective and scaled. That this is so presents no special problem for Aristotle. A man who knows what a saw is, and can readily specify its function, will recognize immediately that a wooden saw is a discretely homonymous saw, and he will not be seduced into thinking that it is a genuine saw. But a man who has only seen a saw on one occasion, when it was used to cut bread because no knife was handy, might still be deceived into thinking that wooden saws counted as genuine saws. Generalizing, then, Aristotle offers as seductive discretely homonymous Fs items which are thought to be real Fs by those who have not yet grasped the *ergon* of Fs, or who have failed to recognize that discretely homonymous Fs cannot perform this function.

Similarly, Aristotle thinks that a fair number of associated homonyms are seductive. These homonyms very often require special pleading. They go unnoticed by common sense; and sometimes philosophers, not beholden to common sense, think of them as univocal. Most people fail to see the homonymy of justice; in the *Republic* Plato thinks he can offer a perfectly univocal account of it (443c–444a). If Aristotle is right, they have equally been seduced into assuming synonymy, one through oversight and the other through excessive reliance on an unwarranted univocity assumption.

I. 8 TYPES OF HOMONYMY: AN OVERVIEW

Aristotle accepts CH as a general account of homonymy. CH recognizes both associated and discrete homonyms as cases of homonymy. Although he sometimes suggests that all discrete homonyms are obviously or even laughably homonymous, Aristotle nevertheless concedes that some discrete homonyms may escape our notice as homonyms. Since their accounts are altogether distinct, FD entails that they are discrete; even so, it is no mere linguistic accident that these homonyms are equally called '*F*'. On the contrary, linguistic custom draws upon entrenched practice to obscure the distinctions FD demands. These are discrete, non-accidental homonyms.

Discrete, non-accidental homonyms are like associated homonyms in tending to escape our notice. They are equally seductive, although seductiveness will in part be a function of the experience of the observer. Mixing the objective, account-driven distinction among homonyms, and the partially subjective demarcation along lines of SH, we find the following types of homonyms:

1. Non-seductive discrete homonyms: keys and the puns of *Topics* i. 15.

2. Seductive discrete homonyms: non-accidental discrete homonyms, including the organic and non-organic bodies.

3. Non-seductive associated homonyms, which may be core-dependent or not: Aristotle evidently thinks of health as in this category, since he uses its homonymy as a model for explicating seductive associated homonyms.[42]

4. Seductive associated homonyms, which may be core-dependent or not: being, goodness, justice, and other key philosophical concepts.

Aristotle draws our philosophical attention especially to (2) and (4). We have thus far seen why (2) carves out homonyms of significance. Of far greater significance to Aristotle are the homonyms captured by (4).

1.9 CONCLUSIONS

Aristotle marks out different forms of homonymy. In doing so, he consistently contrasts homonymy and synonymy in ways which show that he

---

[42] Recall that seductive homonyms can be both subjective and scaled. This does not preclude our marking off a category of philosophical interest, since, Aristotle thinks, key philosophical concepts, e.g. being and goodness, tend to be seductive even to experienced philosophers. Indeed, Aristotle recommends that the subject-relativity of homonymy be exploited for eristic purposes (*Top.* 139^{b}24–9).

regards this distinction as exhaustive. There is, therefore, no *tertium quid* between homonymy and synonymy. Although he sometimes seems to recognize cases of multivocals which are neither homonymous nor synonymous, all of the relevant examples turn out to be cases of homonymy. That they are cases of homonymy is significant for two reasons: (i) if they are cases of homonymy, they do not fall between homonymy and synonymy and so do not threaten CH; (ii) if they are homonymous and not synonymous, we are entitled to follow Aristotle's dominant practice of treating homonymy and multivocity as co-extensive.

Further, when contrasting multivocals and homonyms, Aristotle usually intends to point out merely that some multivocals are not discrete homonyms. He does so to draw attention to the role he envisages for associated homonymy in science, as well as to draw attention to the lapses in syllogistic which result from failing to attend to homonymy. He often illustrates such lapses by relying on discrete homonyms which are immediately apparent to any competent speaker. At the same time, he reserves a class of discrete homonyms for special philosophical attention, precisely because they are not obvious. These are seductive homonyms which are nevertheless discrete. In arguing for their discreteness, Aristotle appeals to FD, the principle of functional determination. Although in fact not perfectly general, FD is sufficient for differentiating some *F*s into discrete kinds. It is, therefore, sufficient to establish non-univocity, and so sufficient to generate homonymy.

FD is, however, impotent when it comes to establishing association. It can show that some *F*s are of the same functionally determined kind; and it can be used to show that some *F*s are not members of the same functionally determined kind. But it cannot be used to show that some *F* things, though distinct, are also associated. Nor can FD suffice even to establish all cases of non-univocity. For not all kinds are functionally determined kinds. Aristotle therefore requires additional mechanisms for establishing univocity and non-univocity; he also requires a clear grounding for association among non-univocals when it exists; and he requires, finally, a way of establishing core-dependence when it obtains.

I turn in Chapter 2 to some of the tests Aristotle relies upon for establishing non-univocity, association, and core-dependence. I ask whether one ought to be persuaded by the kinds of claims he makes on homonymy's behalf.

2

The Promises and Problems of Homonymy

Having established the general framework of Aristotle's approach to homonymy, I turn to a consideration of his reasons for believing that we should take his various appeals to homonymy seriously. We have seen that Aristotle recognizes both discrete homonyms and associated homonyms and that, crucially, among the latter, he sees a special philosophical role for core-dependent homonyms. We have not seen, however, that he is right to detect homonymy as often as he does; nor have we seen that any given concept of philosophical significance actually behaves in the way Aristotle maintains. So, we do not yet know whether Aristotle's various appeals to homonymy should sway us.

One way of beginning to assess the force of Aristotle's claims involves reflecting on the uses to which he puts homonymy. There are two principal uses, one critical and one constructive. These uses correspond in a fairly direct way to the two kinds of homonymy we have identified. On the critical side, Aristotle needs to show that his predecessors go wrong by assuming univocity when a term is in fact equivocal; on the constructive side, he needs to show that a term, though equivocal, is nevertheless not discrete, but associated, perhaps in a core-dependent way. I begin by considering Aristotle's critical and constructive uses of homonymy. I then turn to the sorts of tests for establishing non-univocity upon which Aristotle relies. Next, I reflect more deeply on what is required for establishing association and core-dependence. I do not claim in any general way in this chapter that Aristotle's methods for establishing non-univocity and association fail or succeed; indeed, we must determine success or failure on a case-by-case basis. Instead, I sketch Aristotle's general methods, together with the sorts of problems they invite. These problems, as I conceive them, are the sorts of objections likely to be mounted by a Platonist who is unprepared to take Aristotle's appeals at face value. The degree to which we should take Aristotle's appeals to homonymy seriously depends upon his ability to meet these sorts of objections.

2.1 CRITICAL USE OF HOMONYMY

Aristotle's works are replete with criticisms of his predecessors. His criticisms are often graciously offered; but sometimes they are trenchant or even scathing. He complains on a regular basis, for example, that Plato has failed to appreciate the homonymy or multivocity of a given philosophical term. Thus, an appeal to homonymy frames his several objections to Plato's account of the good: 'since the good is spoken of in as many ways as being is . . . it is clear that there is nothing common and universal in all cases (*EN* 1096ª23–8, with omission). 'For', he continues, 'it would not then have been spoken of in all the categories, but in one only' (*EN* 1096ª28–9). Minimally, in such contexts, Aristotle seeks to convict his predecessors of an undue reliance upon a univocity assumption, the assumption that an analysis of a given philosophical notion will uncover a unified account, expressible in terms of a non-disjunctive definition. In such critical contexts, then, homonymy is principally a destructive tool.

Unsurprisingly, Aristotle highlights the critical applications of homonymy most clearly and self-consciously when offering guidelines for refutation in his rule books on dialectic and eristic, the *Topics* and *Sophistical Refutations*. He repeatedly advises those seeking refutations to uncover instances of homonymy buried in the accounts of their opponents. In such cases, homonymy is usually discrete homonymy and may be as plain and as simple as insignificant equivocity or ambiguity.[1] Thus, homonymy may be no subtler than the ambiguity of 'bank' in:

(1) Juliette deposited her cash in the bank.

and

(2) Ramon took his family to the Seine for a picnic on the bank.

The definitions of 'bank' in (1) and (2) have nothing in common. Thus, no univocal account of 'bank' across these applications is forthcoming. If one's opponent had supposed otherwise, the mere recognition of non-univocity would suffice for refutation.

Because its equivocity is so close to the surface, no one will suppose that

---

[1] I use 'equivocity' somewhat stipulatively of words to indicate that a term has more than one meaning, even when those meanings might be connected in various ways. By contrast, 'ambiguity' denotes a term with more than one meaning, where those meanings are not connected in any semantically interesting way. Thus, every ambiguous term is equivocal, but not every equivocal term is ambiguous. I sometimes refer to ambiguous words as *merely equivocal*, in order to highlight that some equivocal terms may nevertheless be systematically related.

'bank' is univocal. Still, Aristotle believes that there is a form of homonymy akin to the discontinuous equivocity of 'bank' which infects a wide range of philosophical accounts. The cautions expressed in the *Topics* reflect a presumption on Aristotle's part that philosophers habitually overlook even fairly straightforward instances of homonymy. In such cases, a philosopher assumes univocity and so offers an account which respects this as a constraint. Consequently, she offers an analysis which is radically oversimplified.

In these more interesting cases, an unsupportable univocity assumption, while false, will not be witless. Indeed, in some cases, a critic will have to work hard to establish multiplicity. Consider the following:

(3) Bartholomew and Rodrigo are friends.

(4) August and Paula are friends.

Surely, the instances of friends in (3) and (4) may be univocal; but they need not be. The contexts of the utterance may suggest otherwise, that we treat them as non-univocal. If Bartholomew and Rodrigo are business associates, but share no common interest or hobbies and spend little or no time together outside their business dealings, we will appropriately call them friends, but only because of their shared commercial concerns. Since they have had mutually beneficial transactions through the years, they have come to regard each other with special consideration. By contrast, the August and Paula of (4) have deep mutual fondness for each other. They have shared personal interests for decades. More important, they care for one another, each for the sake of the other. August wishes Paula to flourish merely because he is concerned for her own well-being, not because he stands to gain in any way because of her personal successes. Paula reciprocates. She cares for August for his own sake and will feel wounded herself when he suffers. Reflecting on the friendships of (3) and (4), even only briefly, we may feel inclined to think that 'friend' is homonymous. We find ourselves unable to provide a univocal account and disinclined even to try. Still, we may not have noticed earlier that we had used the term homonymously in past applications, because we have never had occasion to discriminate the varieties of friendship.

Moreover, we might usher ourselves into error by assuming that 'friendship' is synonymous. Focusing on (3), we might provide a provisional account of friendship as a mutually beneficial association. We might then be misled to assume that all friendships *essentially* involve mutual benefit. If we had done so in a philosophical disputation, we might have run afoul of some canon of reason not merely by falling into silly ambiguity (e.g. the end is the purpose; death is the end of life; hence, death is the purpose of life), but by relying illicitly on a univocity assumption we later find reason to question (e.g. friendship is a mutually beneficial association; those who are incapacitated cannot benefit anyone; hence, those who are incapacitated cannot have friends).

The most striking instances of this comparatively subtle form of homonymy are also the most difficult to detect. Indeed, their detection will require methods which cannot be appreciated without first understanding the features of Aristotle's metaphysics and epistemology which undergird them. Of special interest in this study will be 'body' and 'oneness'. Aristotle cannot expect a random proficient speaker of a natural language to appreciate that 'body' is homonymous in:

(5) Phillip tones his body by working out in the gymnasium.

and

(6) Lenin's body lay in state for nearly sixty years.

Yet this is evidently just what he means to assert when insisting that a dead body is not a body *except* homonymously (*An.* 412^{b}20–7). The reasons for holding that body as between (5) and (6) is homonymous are hardly obvious, if indeed they are ultimately satisfactory. Presumably, judging by the recommendations of his fullest and most explicit discussion of the sources of homonymy, *Topics* i. 15,[2] Aristotle believes that faulty inferences may result from failing to notice this instance of homonymy. For example, he evidently thinks this inference is mistaken: (i) a human body is potentially alive; (ii) a corpse is a human body; (iii) hence, a corpse is potentially alive. The conclusion may seem initially jarring, if not simply false. If it is to be rejected in the end, then Aristotle will be correct to insist on the homonymy of the body. At any rate, if we assume, as seems initially reasonable, that both premises are true even though the conclusion is false, then Aristotle will be right to insist that 'body' is non-univocal in (i) and (ii).

If eventually they must be accepted,[3] such examples of homonymy as appear in (5) and (6) will show that even unobtrusive instances of the univocity assumption may result in difficulties for those who accept them unreflectively. Of course, it may turn out that these difficulties will be exposed only in the context of philosophical or scientific theorizing, and that in everyday situations, non-univocity may be safely overlooked. Even so, in theoretical situations requiring uncommon precision, a critic who is

---

[2] On *Top.* i. 15, see 2. 4.
[3] I consider the alleged homonymy of the body in detail below, in Chapter 5.

sensitive to homonymy will be in a position to undermine an opponent's unwitting reliance on an unsustainable presupposition of univocity.

2. 2 CONSTRUCTIVE USES OF HOMONYMY

Aristotle's uses of homonymy are not always critical or destructive. On the contrary, homonymy finds equal employment in constructive contexts, including some at the centre of Aristotle's mature positive philosophical theorizing. Certainly he wishes to avoid what he takes to be the mistaken presuppositions of his predecessors. So he wants his own accounts to reflect non-univocity where appropriate. This is why he encourages philosophers to begin philosophical inquiries by trying to detect homonymy (*Phys.* 185ª21–6).

If we focus exclusively on critical contexts, we miss something of central importance to Aristotle's conception of homonymy, and furthermore to his attitudes towards its role in philosophy. For, importantly, Aristotle assumes that in some cases a recognition of non-univocity does not force a commitment to *mere* equivocity. That is, he denies that one can infer rank ambiguity from non-univocity; he contends that these are not exhaustive alternatives. He thinks, instead, that although non-univocal, some philosophical concepts exhibit a kind of order in multiplicity. If this is so, then in his own constructive theorizing Aristotle will seek to mark and display this ordering. This is the ordering he locates in associated homonymy and especially in core-dependent homonymy.

Consequently, Aristotle's reliance upon homonymy in critical contexts as often as not turns out to be destructive only in the sense of preparing the way for a competing positive proposal. That is, he very often claims to find non-univocity in some concept not in order to establish that no unified account of it is forthcoming. On the contrary, having established non-univocity, Aristotle often wishes to find unity within multicity.

This comes out in two ways. First, as we have seen,[4] Aristotle distinguishes the class of good things from a class of homonyms which are, as he says, 'from chance' (*apo tuchês*; *EN* 1096ᵇ26–7). He thereby also implicitly distinguishes a class which are not by chance, which are regulated by non-arbitrary connections. Aristotle does not say what sort of chance he has in mind. It would be natural, but mistaken, to suppose that all discrete homonyms are homonyms by chance; for there is a good explanation, rooted in linguistic custom, why some discrete homonyms have the same names. An axe that cannot function is an axe because of its having once

[4] See 1. 2.

had a function necessary and sufficient for qualifying as an axe. Now it is functionally hollow, even though it maintains the outward appearance of an axe. So chance homonyms seem rather to be, like 'bank', instances of straightforward ambiguity. Consequently, non-chance homonyms may include some instances of discrete homonymy, but they will also include other sorts of cases. For Aristotle has 'goodness' in mind as a contrast with chance homonyms; yet it is not a discrete, non-accidental homonym. The variety of good things, Aristotle rightly assumes, do not all qualify as good in the way that a non-functioning axe qualifies as an axe or a dead body qualifies as a body. These are ex-axes and ex-bodies. Good decisions and good meals do not bear any such relation to one another. So non-chance homonyms will often be associated in ways beyond the ways in which non-accidental discrete homonyms are related.

Second, and more important, Aristotle advertises core-dependence as an especially noteworthy form of association. When he introduces the homonymy of being in *Metaphysics* iv. 1–2, Aristotle illustrates the ways in which various beings are related to one another. The examples he chooses, 'health' and 'medical' (*Met.* 1003ª34–ᵇ4), are supposed to illustrate that healthy complexions and regimens, and medical implements and books, are called healthy and medical because of their standing in some dependency relation to health and medicine. A book is a medical book precisely because it contains information about the medical craft and not some other; the medical craft, by contrast, would remain what it is even if there were no books to describe it. His examples suggest, then, that Aristotle expects core-dependent homonyms to be closely associated and to display discernible forms of dependency. So if he is right that some central philosophical concepts are homonymous in this way, Aristotle must establish not only their association but also their ordered structure.

Taking this all together, we see that in the case of core-dependent homonyms, Aristotle has a three-part task. He must first establish non-univocity. This much is at any rate true of all homonyms. Additionally, though, he must establish some form of association: for without association, positive theorizing will be otiose. That is, without establishing some unity in non-univocity, Aristotle will be susceptible to criticisms discomfitingly similar to those he had launched against his predecessors: he will have assumed a form of unity unsustained by scrupulous examination. His use of homonymy in positive theorizing shows that he is often after more than mere non-univocity. It consequently also places a burden upon him to move to a second stage of argumentation when seeking to demonstrate associated homonymy: having established non-univocity, Aristotle must then display some non-arbitrary form of connection or unity. Finally, given the prominence of his concern with core-dependent homonymy, the form of non-arbitrary unity must have a special structure: every non-core instance must bear some asymmetric dependence relation on the core. So, as a further task,[5] Aristotle must establish and display the focal structure of core-dependent homonyms.

2. 3 HOMONYMY'S FIRST PROBLEMS: POLEMICAL CONTEXTS

Critical and constructive uses of homonymy equally reflect a commitment to non-univocity. A term is homonymous only if it is non-univocal. In the easiest cases Aristotle offers, non-univocity is plain. No one will deny that 'crane' is homonymous as applied to birds and construction machines; and no one will argue that since cranes are animals, the machine parked at the construction site must be an animal. This is because the easiest cases of homonymy are instances of rank equivocity or ambiguity. The easiest cases are always instances of discrete homonymy. Still, as we have seen, not every instance of discrete homonymy will be easy to detect.

At the other end of the spectrum, hardly anyone will grant, at least initially, that 'being' is non-univocal. To begin, there is certainly no patent or undeniable non-univocity between

(11) Socrates exists.

and

(12) Whiteness exists.

On the contrary, one would assume univocity in (11) and (12) without some special reason to avoid doing so. Thus, for example, when denying the existence of universals, a nominalist presumably withholds to whiteness precisely what she affirms of Socrates. As such, she presumes univocity rather than non-univocity. Thus, she implicitly assumes that being is non-homonymous, as least where (11) and (12) are concerned. By contrast, when he maintains that being is homonymous, Aristotle claims to detect a form of multiplicity so subtle that it at first escapes our notice. Indeed, so subtle is this non-univocity that it may well escape even those most sympathetic to the programme of unearthing homonymy wherever

---

[5] This task is only conceptually and not necessarily temporally 'further' than the task of establishing association. One way to establish association would be to establish core-dependence. Still, it should be possible to establish association without establishing core-dependence. Because Aristotle displays little interest in doing so, I do not consider the ways in which he might establish association short of core-dependence.

it lies. A critic who denies that being is homonymous may of course regard Aristotle's claims regarding equivocity as unfounded.

Some of these issues become more sharply focused when we consider the positions of those who self-consciously promote univocity in the very cases where Aristotle claims to find homonymy. Those who are generally unsympathetic to Aristotle's claims, those who continue to insist that core philosophical terms admit of unified, non-disjunctive definitions, may well be unmoved by Aristotle's position. For example, a Platonist who holds that being or goodness is univocal need not—indeed, should not— immediately capitulate to Aristotle's demand that non-univocity be recognized; and a polemical Platonist will rightly demand proof of non-univocity in such contexts. In these sorts of polemical contexts, Aristotle therefore has an obligation to offer compelling evidence of non-univocity; if he fails to offer such evidence, his candidates for homonymy will meet the approval only of those antecedently disposed to find it. Polemical contexts, then, challenge homonymy by demanding tests for non-univocity.

2.4 TESTS FOR HOMONYMY: *TOPICS* i. 15

Aristotle fully appreciates the polemical nature of some candidates for homonymy. Indeed, he sometimes self-consciously upbraids his adversaries for failing to recognize instances of homonymy, even when the proposed instance is far from uncontroversial. Thus, in *Nicomachean Ethics* i. 6, Aristotle claims that those who introduced the Forms, the Platonists,[6] failed to appreciate the priority and posteriority of goodness in various categories. Given that substances, quantities, and qualities are all called good, and given that substances are 'prior in nature' (*proteron tê(i) phusei*) to quantities and qualities, 'there could not be something common, an Idea <set over all> these things' (*EN* 1096ª21–3).

Minimally, Aristotle's argument here involves him in denying the univocity of 'good' in:

(13) Xanthippe is good.

(14) Virtue is good.

(15) A stock portfolio weighted towards technology firms is good.

---

[6] Aristotle does not mention the Platonists by name in this connection; but it is overwhelmingly likely from the content of the discussion that these are his intended targets. See also *EN* 1172ª5, 1177ª34.

This, it seems, is the purport of Aristotle's claims that there cannot be a common Idea or Form set over all goods and that the good 'cannot be universally present in all cases and single'. Aristotle does not, however, regard this instance of homonymy as clear. 'At any rate,' he says, '<goodness> does not resemble those <homonyms which are> homonyms by chance' (*ou gar eoike tois ge apo tuchês homônumois*) (*EN* 1096ᵇ26–7).

Suppose, then, that I am a Platonist who believes that goodness is a simple, non-natural indefinable property.[7] Suppose further that I believe that Xanthippe is good because she participates in Goodness and that the same holds true of virtue and moderation.[8] I cannot be expected at this point simply to grasp the non-univocity of 'goodness'. If Aristotle has nothing to substantiate his contention, we will have, at best, a stalemate. Moreover, since he introduces the non-univocity of goodness in the context of criticizing a position I have articulated, I may justifiably demand argument in support of his conclusion at this juncture.

Aristotle does not shy away from the task of providing arguments for homonymy. In *Topics* i. 15, he offers a series of *homonymy indicators*, each in its own way devised to be a test for non-univocity.[9] These include:

> (i) a test for the forms of contrariety terms enjoy (since the contrary of 'sharp' as applied to music is 'flat', but as applied to intelligent persons is 'dull', 'sharp' is homonymous) (106ᵃ9–21);

---

[7] Suppose, for example, I hold the view articulated and defended in Butchvarov 1982; he argues that goodness is a simple, non-natural property and that any attempt to define it will fail.

[8] Holding such a view does not commit the polemical Platonist to the manifestly false view that there is one Form answering to every general term. The Platonist can certainly recognize ambiguity as well as anyone else. Rather, the claim now introduced is the more restricted thesis that there is only one property, Goodness, and that everything which counts as good does so because of its participating in this Form. Aristotle is sometimes credited with showing that Plato is wrong to hold that there is one Form corresponding to every general term. Surely Plato never held such a transparently false view. See Fine 1993, together with Kraut's review (1995). That said, I am sympathetic to Kraut's observation (1995, 117) that 'Plato's confidence that there are forms of large and bed must owe something to his assurance that much of the classificatory work done by ordinary general terms is philosophically defensible. If he had lacked this confidence, then he would have held that what names there are has no relevance to the determination of which forms there are, and in this case it is unlikely that he would have appealed to the "many things to which we apply the same name" in his formulation of the one over many argument. If what matters to this argument for forms is the unity of the many . . ., then he has no business appealing to names here unless he takes names to be at least a rough guide to what unities there are. If he had thought that for the most part what unities there are remain unnamed, and that what names we have typically do not pick out unities, then the things "to which we apply the same name" would have been of little philosophical interest to him.' Kraut's attitude towards Plato captures exactly mine towards Aristotle.

[9] Some of Aristotle's tests are unclear; others do not translate readily into English. For the purposes of this exposition, I have adapted them freely.

(ii) a test for the existence of contraries (love, the emotion, has hate as its contrary while physical love lacks a contrary altogether) (106^a24-35);

(iii) a test for intermediates (black and white issues are opposed to grey areas, with no clear intermediaries, while black and white, the colours, are opposites and have grey and the whole of the colour spectrum between them) (106^a35-^b12);

(iv) a test based upon difference in contradictory opposites (failing to perceive is in one case contradicted by sensing and in another by grasping the point) (106^b14-20);

(v) a test based upon inflections and paronomy (if 'judiciously' is applied differently to a magistrate's way of ruling and a billiard player's strategizing, then 'judicious' will likely be homonymous) ($106^b29-107^a1$);

(vi) a test determined by a term's signification ('clear' as applied to sheets of glass signifies transparency, while 'clear' as applied to consciences signifies freedom from guilt) (107^a3-18);

(vii) a test for sameness of genus (some cranes belong to the genus of animal, while other cranes are machines) (107^a18-30);

(viii) a test based on definition and abstraction (defining 'bright' in 'bright girl' yields 'intelligent girl', whereas 'bright' in 'bright light' yields 'shining light'; abstracting away the modified terms, we have 'intelligent' and 'shining'; this difference would not present itself for univocal terms) (107^a36-^b5);

(ix) a test of comparability (a knife is sharp and so is a professor; but the professor cannot be sharper than the knife; since univocal terms are comparable, 'sharp' is non-univocal in these applications) (107^b13-18);

(x) a test based on the differentiae of genera which are not subordinate or superordinate to one another ('flat' differentiates one kind of sound from another, and also one kind of terrain from another; since sounds and terrains are not genera related by subordination or superordination, 'flat' is non-univocal) (107^b19-26);

(xi) a test based on distinctness of differentia (flowing symphonies do not have the same differentia as flowing rivers) (107^b26-31); and

(xii) a test to see if one term is used once as a differentia and another time as a species, since a species is never also a differentia (blue is species of colour but also differentiates one kind of mood from another; so 'blue' must be non-univocal) (107^b32–6).

Passing at least one of these tests is presumably sufficient for establishing homonymy, because any one of them will show that a term is non-univocal. Moreover, passing at least one is necessary but not sufficient for establishing associated homonymy, because a term may be equivocal without being an associated homonym. Moreover, with one exception,[10] none of these tests is even necessary for establishing non-univocity; many of Aristotle's preferred homonyms do not pass them.

Significantly, 'goodness' does not pass them all; nor does it pass even one of them in a striking or compelling way. Certainly a Platonist could, for example, simply maintain that the contraries of (13)–(15) are:

(16) Xanthippe is bad.

(17) Virtue is bad.

and

(18) A stock portfolio weighted towards technology firms is bad.

In this case, we do not obtain the result that we obtain when offering the contraries of

(19) Her *daube de bœuf béarnaise* is perhaps a bit rich.

and

(20) These, then, are the foibles of the rich.

as

(21) Her *daube de bœuf béarnaise* is perhaps a bit bland.

and

(22) These, then, are the foibles of the poor.

Perhaps this is to be expected. For, in fact, we need not have performed the recommended tests to determine the homonymy of 'rich'. Since we appreciate it straightaway, the tests suggested are mostly superfluous. In the same way, it is hardly necessary to see a large bank account will not make one richer than *daube de bœuf béarnaise* in order to see that 'rich' is

---

[10] The exception is (vi), a test based upon signification. As I make clear below in 2. 5 and in 3. 3, this is by far the most important and difficult of Aristotle's tests for homonymy.

non-univocal in (19) and (20). By contrast, when homonymy is not evident, the recommended tests prove much less than decisive.

This, then, is the first problem with Aristotle's apparatus of homonymy: he must establish non-univocity in polemical contexts. In the most controversial cases, however, his standard homonymy indicators will not yield uncontroversial results.

2.5 TESTS FOR HOMONYMY: SIGNIFICATION

Aristotle's sixth test for homonymy merits special attention, because it raises a fundamental question about the way Aristotle understands and applies homonymy. When Aristotle speaks of homonymy and multivocity, we naturally take him to be offering an observation about words. As a rough starting point, so far we have been inclined to regard homonymous words as akin to ambiguous words, although with two reservations. First, Aristotle's doctrine encourages us to recognize ambiguities we might otherwise overlook; and second, the ambiguities thus uncovered may nevertheless display systematic interconnections of philosophical interest.

Topics i. 15 introduces signification as a test for homonymy in a way which evidently supports our tendency to regard it as primarily a linguistic doctrine, or at least as one which finds its most natural expression as a doctrine concerning the meanings of words. It is, to begin, introduced alongside a series of other tests which make free appeal to linguistic intuition:

It is necessary also to consider the types of predicates <signified> by the name, <to determine> whether it is the same in all cases.[11] For if it is not the same, it is clear that what is said is homonymous. For example, in the case of eating, the good is what is productive of pleasure; in the case of medicine it is what is productive of health; in those cases where it is said of the soul, it is a quality, e.g., temperate or courageous or just—and it is similarly applied to man. Sometimes it is <predicated> of a time, as e.g. the good in a propitious moment, for what happens in a propitious moment is called good. Often it is <predicated> of a quantity, as e.g. when it is applied to the mean, for the mean is called good. The result is that the good is homonymous. (*Top.* 107ª3–12)

The argument here seems simple, straightforward, and telling. If '*F*' in '*a*

---

[11] (i) Reading *de dei* with C. (ii) Aristotle does not use *sêmainein* in this passage. The context strongly warrants the judgement that Aristotle is relying on conditions of signification. For he treats the categories as things signified (*Cat.* 3ᵇ10–23) and here introduces them as what falls under a name. For this reason, the Revised Oxford Translation (Barnes 1984) is right to begin rendering this passage as 'Look also at the classes of the predicates signified by the term . . .'.

is *F*' and '*F*' in '*b* is *F*' signify different things, then '*F*-ness' is homonymous. So, difference in signification is sufficient for homonymy. We might add: since what a word signifies is its meaning, difference in meaning is sufficient for homonymy. Understanding signification this way, we will expect Aristotle to point to difference in meaning when attempting to establish non-univocity; and we further expect him to introduce words as the bearers of meanings in these contexts.

Aristotle sometimes meets this expectation. As we have already noted, he sometimes says directly that words are homonymous (*GC* 322ᵇ29–32). Other passages also have nearly this purport, including most notably *Topics* 106ᵇ29–107ᵃ2, where Aristotle talks about the cases or inflections of homonymous items: 'whenever it <the original term> is used in more than one way, an inflection (*ptôsis*) of it will also be said in more than one way; and if the inflection is used in more than one way, so too is it <the original term>' (106ᵇ37–107ᵃ2). He uses examples like 'health' (*hugieia*), 'healthy' (*hugieinon*), and 'healthily' (*hugieinôs*), where the items under description are evidently words.

We may on the other hand emphasize the singularity of the passage. Perhaps it should strike us as odd that this is the only passage in the entire corpus where words are called homonymous. If he believes that words are homonymous in a way akin to the way words are ambiguous, we should expect him to note with some frequency that homonymous words have distinct senses, or that the senses of homonymous words are related in certain ways.

Aristotle does not meet this expectation. On the contrary, he quite regularly talks of *entities* as being homonyms. Thus, he standardly says that an entity of a certain kind *F* that loses the capacity definitive of being an *F* is no longer an *F*, except homonymously. For example, in illustrating a point about the relation of the soul to the body, Aristotle mentions the eye and its relation to sight: '[I]f the eye were an animal, sight would be its soul, for this is the being of the eye with respect to its account. The eye is the matter of sight; if sight is lacking, it is no longer an eye, except homonymously, just as a stone eye or a painted eye <are no longer eyes, except homonymously>' (*An.* 412ᵇ18–22). In applying the doctrine of homonymy in this way, Aristotle does not remark on the senses the word 'eye' has; nor does it seem appropriate that he should. He says that human eyes which cannot see are not eyes at all except homonymously. They are more like the eyes one finds in paintings and sculptures than they are like the fully functioning eyes of living animals. Moreover, he provides a reason for thinking this way. When an *F* has lost what is definitive of being an *F*, it is no longer an *F* as such. This appeal to the being of a kind with respect to its account (*ousia ophthalmou hê kata ton logon*) seems far removed from any appeal to the senses a word may have. The illustration serves to show not that 'eye' has several meanings, that 'eye' is ambiguous in some way, but rather that nothing is a genuine eye unless it can perform the function eyes characteristically perform, namely, seeing. This is in part what Aristotle means in appealing to being with respect to the account.

Some may, then, wish to contend that Aristotle's claim regarding the homonymy of words at *De Generatione et Corruptione* 322^{b}29 as relaxed or anomalous. His more prominent view treats entities, not words, as homonyms, and treats homonymy not as a semantic phenomenon, like ambiguity, but as a metaphysical principle concerned with the substances or essences of kinds.

However that may be, Aristotle's procedures appear inconsistent. Consequently, we have a second problem about homonymy: its domain is not fixed. That is, it is unclear whether homonymy pertains to words and their senses or to entities and their essences. It is therefore also unclear whether it is appropriate to appeal to difference in meaning when trying to establish homonymy. In particular, it is at this stage unclear whether it is appropriate to appeal to linguistic intuition when disputing about whether something, e.g. goodness is homonymous. For it is unclear whether difference in meaning is necessary or sufficient for non-univocity.

Before determining how we should deal with this problem, we will need to reflect upon Aristotle's primary philosophical motivation for introducing the doctrine in the first place. We shall also have to reflect upon Aristotle's approach to signification; for if signification seems initially to be a meaning relation, then since difference in signification is sufficient for non-univocity, difference in meaning will be sufficient for homonymy. After considering his approach to these matters, we shall be in a position to determine whether the problem thus mooted is a genuine one for Aristotle or whether it in the end involves a false presupposition.[12]

---

[12] Some will respond in a deflationary vein. It is perhaps conceivable that Aristotle is guilty of a low-level use-mention confusion. If his official doctrine turns out to be that things, and not words, are homonymous, it may be that he simply slides too readily into calling the words which designate homonyms 'homonymous'. Thus, if banks are homonyms, then Aristotle may remark that 'bank' is homonymous, where this is in the end no more than an abbreviated way of saying that those things designated by 'bank' are homonyms. This would not be remarkable. We speak of the 'subject' of a sentence sometimes in terms of a noun phrase, and sometimes in terms of the entity designated by the noun-phrase; we may even move back and forth in the same context. ('The subject of the sentence in English normally stands in the first position, and in an active-voice sentence performs the action designated by the verb.') Although we are strictly guilty of use-mention confusions in such cases, no serious misunderstanding results. Even so, this response merely postpones the very real question of whether difference in meaning is necessary or sufficient for homonymy.

As a working rule, I will follow Aristotle's own practice by speaking indifferently about words and things as homonyms, even where I raise the question of whether homonymy per-

2. 6 FOCAL MEANING, FOCAL CONNECTION, AND CORE-DEPENDENCE

Questions about the character of Aristotelian homonymy turn crucially on the nature of signification. If signification is a linguistic notion, then difference in meaning will be sufficient for homonymy. If it is an entirely non-linguistic notion, then difference in linguistic meaning will be largely irrelevant to establishing non-univocity. Since every homonym involves non-univocity, every instance of homonymy, even those deployed in purely destructive contexts, will involve difference in signification, however it is to be understood.

The nature of signification also figures prominently in determining whether we ought to accept Aristotle's applications of homonymy in constructive contexts. As we have seen,[13] if we judge by Aristotle's favourite examples, instances of homonymy in constructive contexts involve both non-univocity and co-ordination. Recall that 'health' is homonymous in propositions (7)–(9) above (Socrates, his complexion, and his regimen all count as healthy), even though the various applications of the predicate are clearly somehow related, whereas the applications of 'bank' in (1) and (2) are not. If signification is a linguistic notion, and it is nevertheless an arbiter of non-univocity (*Top.* 107ᵃ3–12, together with the other homonymy indicators of *Top.* i. 15), then presumably the connections we recognize in (7)–(9) but deny in (1) and (2) will themselves be linguistic. That is, we may be inclined to say, in a non-technical way, that the meanings of 'healthy' are related to one another, whereas the meanings of 'bank' in (1) and (2) are not. If so, the form of co-ordination we recognize will be at least partly determined by our linguistic intuitions. Aristotle will then be in a position to appeal to such intuitions when arguing for both non-univocity and co-ordination. If we are independently inclined to regard the relations between the various uses of 'healthy' as broadly linguistic, then this itself will offer evidence that signification concerns meaning. To

---

tains principally to words or things. Sometimes, as context warrants, I will mention words and comment on their homonymy; I will otherwise talk about things as homonyms without recourse to language. If in the end Aristotle is guilty of a benign use-mention confusion, the purport of his doctrine will nevertheless be clear. A more difficult question concerns whether some interpreters have committed a deeper, more malignant confusion on his behalf by explicating homonymy exclusively in terms of the senses of homonymous words.

As an example of commentators who are deflationary on this score, see Williams 1982, 113: 'but it is after all a fact about what grows on palm-trees and what we write at the top of letters that we use the word "date" to describe them both. The fact is only a fact about words in so far as it is also a fact about things, and vice versa.'

[13] Section 1. 6.

this extent, then, our first inclination to treat homonymy as driven in part by differences in meaning will receive further confirmation.

This would be useful for Aristotle, since it is incumbent upon him to specify some form of co-ordination for those homonyms used in constructive theory building. If homonymy and multivocity concern meaning, at least in part, then Aristotle will justifiably rely upon linguistic intuition when determining whether a given non-univocal term is merely equivocal or is, rather, somehow co-ordinated. He will also be justified in relying upon linguistic intuition when determining which application of a predicate is prior to the others.

His own preferred examples may be illustrative. As we have seen, when introducing the science of being qua being in *Metaphysics* iv. 1 and 2, Aristotle offers 'health' as an uncontroversial example of what I have called *core-dependent homonymy*:

Just as everything which is healthy is related to health (*pros hugieian*), some by preserving health, some by producing health, others by being indicative of health, and others by being receptive of health; and as the medical is relative to the medical craft (*pros iatrikên*), for some things are called medical because they possess the medical craft, others because they are well-constituted relative to it, and others by being the function of the medical art—and we shall also discover other things said in ways similar to these—so too is being said in many ways, but always relative to some one source (*pros mian archên*). (1003ᵃ34–ᵇ6)

Crucially, Aristotle commits himself to holding that 'healthy' not only associates, despite its being non-univocal, but associates around some one core or source.

In this sense, his examples are intended to illustrate a general, formal constraint on core-dependent homonymy:

CDH: *a* and *b* are homonymously *F* in a core-dependent way iff: (i) *a* is *F*; (ii) *b* is *F*; and either (iii a) the account of *F* in '*b* is *F*' necessarily makes reference to the account of *F* in '*a* is *F*' in an asymmetrical way, or (iii b) there is some *c* such that the accounts of *F*-ness in '*a* is *F*' and '*b* is *F*' necessarily make reference to the account of *F*-ness in '*c* is *F*' in an asymmetrical way.

As we have seen,[14] 'health' evidently meets this formal constraint.

Still, there is a crucial unclarity in CDH. It concerns the force of the modal claim that one account *necessarily* makes reference to another. If the reason is that the meaning of the non-core application of the predicate can be understood only in terms of the meaning of the core applica-

---

[14] Section 1. 6.

tion, then Aristotle once again will be able to rely on linguistic intuition to adjudicate questions of non-univocity and association. There are obvious advantages to this way of proceeding. Even in polemical contexts, there will be a decision procedure for adjudicating disputes. In order to determine whether 'goodness', for example, is non-univocal and, if so, whether there is a core notion, we can simply appeal to the linguistic intuitions of competent speakers of our language. We will then be able to identify the *focal meaning* of goodness and to show how non-core applications of the predicate relate to it.[15]

There are also obvious disadvantages to this way of thinking of core-dependent homonymy. To begin, linguistic intuitions are apt to come up short in the very contexts in which we have difficulty settling on non-univocity or association. Although perfectly adequate for 'bank', our intuitions about correct applications of the word 'good' will hardly decide the issue of whether goodness admits of a univocal account. Nor, indeed, should linguistic intuition suffice in such contexts. Presumably a dispute between Aristotle and the Platonists about whether goodness is homonymous concerns the question of whether there is some one universal, goodness, common to all good things. This is surely how Aristotle conceives the issue himself: he claims, against the Platonists, that 'the good is not some common thing, universal and one' (*EN* 1096ᵃ27–8). In urging this conclusion Aristotle seems indifferent to matters of linguistic use or intuition. If so, then the core-dependent homonymy he sees in goodness is not focal meaning, but another form of *focal connection*.[16]

If we understand core-dependent homonymy in terms of non-linguistic focal connection, our question about the modal force of the necessity in CDH reasserts itself. To see this most clearly, we can consider two different types of examples, each rather strained but nevertheless sufficient to put pressure on the form of necessity required for association. In the first instance, someone might provide an account of a bank as an institution for saving and trading money within a thousand miles of a sloping side of a river. Suppose further that it is a non-contingent physical fact that every point on Earth is within a thousand miles of a river. (Perhaps this is a consequence of a geological law having to do with the movement of tectonic plates.) In such a case, an account of savings banks will make reference to river banks, and this will not be wholly accidental. Obviously, this cannot suffice for the form of association required for focal

---

[15] The term 'focal meaning' originated in Owen 1960.

[16] The term 'focal connection' originated with Irwin (1981), who self-consciously prefers it as an alternative to Owen's semantic or linguistic conception of homonymy as focal meaning.

connection. As a second example, suppose someone provides an account of healthy regimens as those involving vigorous exercise and balanced diets. This person could grant non-univocity in the case of 'healthy', and could even grant some form of association; but she could deny any form of core-dependence. If we were to point out that the meanings of 'healthy' as applied to persons and regimens are clearly connected, she might or might not agree; but she could insist that linguistic connection is in any case irrelevant and that there would be no reason to presume any form of connection sufficient for core-dependent homonymy. (She might say this, even though she finds other instances persuasive.)

These two cases provide converse sides of the same problem: Aristotle requires co-ordination for core-dependent homonymy. If the forms of association are non-linguistic, then we require a principle for determining association and core-dependence, as well as a method for applying the principle. Since no simple appeal to linguistic connection will be appropriate, we are left without any clear direction. Surely the formal constraints introduced by CDH must be respected; but this does not take us nearly far enough. For CDH requires a modal connection between one account and another; yet CDH alone specifies neither the modal status of this connection nor the substantive form of dependence required.

Each of these cases brings out further related problems. In Aristotle's preferred examples, there is no obvious unity to the kinds of relations non-core homonyms bear to core homonyms. Thus, healthy regimens are *productive* of health; healthy complexions are *indicative* of health; and a healthy tonic is *preservative* of health. These relations have no intrinsic connection with one another; nor do they by themselves determine what additional relations are appropriate; nor again is any one of them sufficient to ensure core-dependence. Human beings are productive of other human beings. But 'human being' is univocal as applied to parents and children.[17] Falling barometers are indicative of changing weather patterns. But 'change' is not applied homonymously to weather patterns and barometers. Further, judging by just those relations Aristotle cites, it is impossible to say directly whether 'being within one thousand miles of' is the right sort of relation for establishing focal meaning or connection. CDH provides some initial help; but so far we have in it only a formal account, and not one which tells us when one account *must* appeal to another. We cannot admit junk relations in determinations of core-dependence; CDH by itself does not provide us a principle for sorting appropriate from junk relations.

[17] 'Animal' is synonymous in 'Socrates is an animal' and 'Pavlov, the dog, is an animal'. See *Cat*. 1ᵃ6–12.

The problem here is that the sample relations Aristotle offers in illustrations of homonymy do not by themselves provide any substantive principle for determining and delimiting relations of association or core-dependence. This is reflected in CDH, which does not promote a method for carving off unwanted, bogus relations of core-dependence. Fairly clearly, no first-order account of the appropriate forms of association will be forthcoming. What is needed, therefore, is a second-order account which provides a framework for core-dependence without enumerating the acceptable relations themselves. Such a second-order account would have the doubly desirable effect of providing adequacy conditions for core-dependence while being open-ended in the sense of permitting the recognition of new relations of core-dependence when appropriate.[18] What is wanted in this context is analogous to what is wanted in an account of 'poison'.[19] In offering an account of poison, we do not wish merely to specify the toxic substances there are, since new ones may yet be discovered or invented. Still, we can give a perfectly determinate second-order account in terms of the characteristic causal effects of poisonous substances. An account of appropriate association will be similarly both determinate and open-ended, though it may or may not be similarly causal.

Aristotle never provides a second-order account of the appropriate forms of core-dependence. Still, I shall argue, his examples suggest a way to specify CDH in an appropriately substantive way, provided that certain broader theses of his metaphysical realism are accepted.[20]

Taking all this together, we have a series of nested problems that all have to do with the focal meaning or connection of core-dependent homonyms. First, and certainly most important, it is necessary to determine whether the appropriate relations are meaning relations or not. Aristotle's standard examples of homonymy are naturally understood in terms of meaning relations. Even so, his more precise and self-conscious

[18] This is what is correct in Owen's suggestion (1960, 189) that homonymy must be open-textured: 'Aristotle has not solved the problem of defining focal meaning fully and exactly so as to give that idea all the philosophical power that he comes to claim for it: he has given only the necessary, not the sufficient, conditions for its use. But there is no reason to think that this problem can have a general answer. Aristotle's evasion of it may come from the conviction that any answer would be artificial, setting boundaries that must be endlessly too wide or too narrow for his changing purposes. The concept of a word as having many senses pointing in many ways to a central sense is a major philosophical achievement; but its scope and power are to be understood by use and not by definition.' Owen evidently misses the possibility of a second-order account which is both determinate and open-ended, in the sense that it provides clear conditions of application but does not enumerate or otherwise delimit the possible range of cases.

[19] Armstrong 1968.

[20] I offer this account below in 4. 4 and 4. 5.

accounts of homonymy do not appeal to meanings in any crucial or in-eliminable way. This problem needs to be investigated by determining whether, and if so, how, signification is for Aristotle a meaning relation. For difference in signification determines homonymy.

Within this broad, foundational problem about the nature of associa-tion for core-dependent homonyms lies a series of subordinate problems. In offering CDH, Aristotle presupposes that we have a clear conception of account dependence; but he provides no clear statement of what this consists in or requires. Consequently, Aristotle provides no clear method for determining associativity or core-dependence in disputed cases. Since homonymy has some of its most important applications in polemical con-texts, Aristotle will need to point to a method which can be accepted by all parties to a dispute. Finally, whatever these methods are to be, core-dependence cannot be so close that it suffices for univocity in the end. That is, there is ever a danger of having falsely assumed non-univocity when offering our initial account. Thus, surveying the many forms of toxins and mechanisms for inducing toxicity, someone might decide that 'poison' is non-univocal. When eventually grasping the appropriate second-order account, he will be forced to see that it is after all perfectly univocal. Perhaps 'goodness' or 'friendship' will turn out this way. Hence, judge-ments of non-univocity must be perfectly secure, but not so secure that they guarantee non-associativity; judgements of association and core-dependence must likewise be firm and determinate, but not so close that they re-introduce an overlooked univocity. Aristotle must, therefore, establish order in multiplicity without surprising himself by uncovering more order than he had initially recognized.

2. 7 HOMONYMY'S PROBLEMS: AN ILLUSTRATION

It is easy to appreciate the difficulty and delicacy of judgements of core-dependent homonymy when considering highly contentious cases like 'being' and 'goodness'. Because they are so abstract, however, such cases do not provide serviceable illustrations of the problems of homonymy. Far better in this regard are mid-range cases, not so polemically charged. One such illustration is 'life'.

Aristotle suggests that 'life' is homonymous:

Let us say, then, in taking up a new starting point for our inquiry, that what is ensouled is distinguished from the unensouled by living. But 'living' is spoken of in many ways, and if even one of these belongs to something, we say that it is alive, that is: thought; perception; motion and rest with respect to place; and further motion with respect to nourishment, decay and growth. For this reason all plants too seem to be alive. (*An.* 413ᵃ20–6)

Accordingly, Aristotle criticizes Dionysius for having offered a univocal account of 'life': Dionysius had thought that life involved a certain kind of native movement in a creature capable of taking on nutrition (*Top.* 148ᵃ26–31).²¹ Evidently, Dionysius mistakenly failed to recognize that life is not the same for plants and animals, and that it is therefore homonymous. Aristotle himself wants to correct this misapprehension in the *De Anima* when he claims that 'life is spoken of in many ways'.

If homonymous, 'life' must be non-univocal. Moreover, if it is a core-dependent homonym, it must not be merely equivocal. Rather, the various instances of life must be co-ordinated in systematic ways. Indeed, according to CDH, some non-core instances of life must, in their accounts, asymmetrically depend upon some core instance. Aristotle must first establish multiplicity. He must then find order within that multiplicity.

Perhaps showing that 'life' is non-univocal strikes us as easy. If we consider the following common sentences about life and being alive, we may simply agree straight away:

Vice-President Gore is alive.
The human race is alive.
This trout is still alive.
My cherry tree is alive.
The entire ecosystem is alive.
(Said of a computer) It's as alive as you and I are.
Of course Santa Claus is alive. If he were dead, how could he deliver all those presents?
Her candidacy remains alive.
It is fair to say that though she is no longer among us, she lives still in our memories.
Despite its unpopular social policies, the Roman Catholic faith is alive and well.
If all viruses are alive, then so is the one screwing up your hard drive.

For human beings, thought is life.
For her, life is meaningless.
This river has witnessed an awful lot in its long, languid life.
Every empire has its life: it is born, flourishes, totters, decays, and dies.
Village life is just too constraining for me.

---

²¹ See 7. 2 for a fuller discussion of Dionysius.

It's a pity—he was so full of life.
This new CEO has a life of two years, maximum.
Maria Callas is my life.
Thereupon, he embarked on his new life.
This is the life!
For many creatures, water is life.

It would be difficult for anyone surveying this list to regard 'life' and 'alive' as univocal across all of its applications. One might, consequently, assume without further reflection that 'life' is non-univocal, in order to begin the chore of determining whether it is nevertheless a core-dependent homonym.

This conclusion would be over hasty. First, our initial evidence for non-univocity in such contexts is linguistic intuition. Surely, it might be thought, 'life' does not have the same meaning when applied to candidacies and children and so must be non-univocal. But if we think that homonymy and signification are not concerned at all with meaning, such intuitions will be idle. Difference in meaning will not establish non-univocity. Second, and much more important, Aristotle allows that a term can be homonymous in some of its applications but not in others (*Top.* vi. 10); and he evidently thinks that 'life' is homonymous across a narrow subset of these applications. For his criticisms of Dionysius presuppose that 'life' is homonymous across even the following applications:

(23) Socrates is alive.

(24) Heather's dog, Cynic, is alive.

(25) My floribunda rose bush is alive.

Here even if we were to admit linguistic intuition as evidence for homonymy, we would be hard pressed to agree with Aristotle. Instead, we might agree that 'life' is homonymous across the broad list, only to insist that it is univocal in (23)–(25).

This illustrates homonymy's first problem. In all but trivial cases, it is difficult to establish non-univocity. It is, moreover, difficult to see immediately how we are to adjudicate contested cases, particularly if we are not to rely on linguistic intuition as a source of evidence.

Suppose, however, that we are in the end persuaded that 'life' is non-univocal in (23)–(25).[22] Further, suppose provisionally that we come to appreciate that we might rewrite (23)–(25) as:

---

[22] I consider the prospects for the non-univocity of life in Chapter 7.

(23′) Socrates engages in rational activity.

(24′) Heather's dog, Cynic, engages in perceptual activity.

(25′) My floribunda rose bush engages in nutritive and photosynthetic activity.

Given the assumption we have made, we may come to agree that (23′)–(25′) contain distinct predicates, and we may conclude that since they are glosses on (23)–(25), these too, upon analysis, are shown to be non-univocal.

Having gained non-univocity, we then turn to establishing association and core-dependence. To do so, in accordance with CDH, we must establish a core upon which non-core instances depend. It is hard to see what Aristotle might take the core to be. Certainly it would be wrong to take the form of life Socrates enjoys as the core. There is no obvious reason to suppose that an account of the life Cynic enjoys must be understood in terms of Socrates' rational activity; still less must the life of a rosebush be explicated by reference to rational activity.

Perhaps the problem here is that the direction of analysis is backwards. Perhaps, instead, we should think that the life of my floribunda rosebush provides the core. Aristotle at any rate thinks that in the mortal realm anything that is rational also engages in nutritive activity (*An.* 423ᵃ25–ᵇ1, 415ᵃ22–6). Even so, it is not possible for Aristotle to accept nutritive activity as the very core of life. For he thinks that god is alive (*Met.* 1072ᵇ29–30), even though god is an immaterial particular form.[23]

This illustrates homonymy's second problem. Having won non-univocity, we are in danger of losing association and core-dependence.

One might think that this problem is easily overcome by the right form of analysis. That is, instead of supposing that some one of the living entities exhibits the core of life in terms of which the lives of the others must be understood, it would be better to focus on Aristotle's express view of the matter. 'By life', he claims, 'we mean nourishment, growth, and diminution through oneself' (*An.* 412ᵃ14–15; cf. 415ᵃ28–ᵇ18). This suggests that instead of treating some one of the instances of life as fundamental, we should treat at least all those surveyed as non-core, depending equally upon something common to them all. Thus, as Aristotle suggests, life involves a specifiable, self-generated end-directedness. Life, then, is not

---

[23] God is a substance and so particular; as a particular substance, god is either (i) matter, (ii) a compound of form and matter, or (iii) a form. God is, however, without matter. Hence, neither (i) nor (ii) is possible for god. Hence, god is a form, an immaterial particular form.

any first-order way of being end-directed (e.g. taking in nutrition, engaging in practical reasoning), but rather acting from an origin within oneself for the purpose of achieving some stable goal. In the case of plants, non-human animals, and humans equally, this will consist in a form of flourishing, both for the individual and for the species of which it is a member. Considered in this way, then, the core of life is self-generated, end-directed activity for the purpose of fully realizing one's nature, that is, for flourishing.

Now, however, we need to rethink whether we granted the non-univocity of 'life' prematurely. Having established some form of unity in multiplicity, why have we not simply uncovered the univocity that had earlier eluded us?[24] This is a special problem in non-trivial cases. For here we do not have any secure pre-theoretic intuition that life, for example, differs across closely related applications, as in (23)–(25). Indeed, we have no obvious set of relations of the sort putative non-core applications bear towards some core, as called for by CDH. For this reason alone, we may wonder whether we have non-univocity at all. Consequently, for this same reason, we may wonder whether we have an instance of homonymy at all.

This, then, is homonymy's third problem. Whenever we move away from the simplest cases, association may prove so close that we need to rethink our initial anti-Platonism.

This third problem can be generalized. Judgements about univocity and non-univocity are influenced by the level at which we make them. That is, when we observe that life is realized in different ways by different organisms, we may become inclined to view 'life' as non-univocal. Yet it may turn out that such first-order judgements are incorrect or otherwise premature. For second-order univocity is consistent with first-order multiple realizability. The breast-stroke, backstroke, and freestyle crawl are ways of swimming; but 'swim' is univocal when predicated of swimmers swimming these different strokes. Perhaps belief is multiply realizable; still, a second-order, functional account of belief may be indisputably univocal. In general, then, first-order multiplicity is not sufficient for non-univocity. Consequently, it will sometimes prove unclear whether Aristotle is guilty of overlooking the resources the supporters of philosophical univocity have at their disposal.

---

[24] So Irwin concludes (1988, 283–4) that life is after all not homonymous for Aristotle: 'This account of life is Aristotle's answer to the suspicion that being alive is really several homonymous properties. He thinks being alive is an important common property because it is a teleological order of the parts and bodily processes of the organism to achieve some constant goals.'

2. 8 HOMONYMY IN ARISTOTLE'S DEVELOPMENT

Whether or not he finally accepts the homonymy of 'life', Aristotle remains constant about its potential multiplicity throughout his career. This is not the case with a second disputed case of homonymy, one which also serves to illustrate the problems involved with establishing non-univocity and association but which raises an independent question about Aristotle's attitudes towards core-dependent homonymy itself.

Early in his career, in the *Eudemian Ethics*,[25] Aristotle introduces friendship as homonymous:

There must then be three kinds of friendship, not all named for one thing or as a species of one genus, nor yet as having the same name quite by mere chance. For all the senses are related to one which is primary. The primary is that whose definition is contained in <the definitions of all> by us. (*EE* 1235ª17–20; cf. 1236ª7–14, ª17–19, ᵇ21–6)[26]

Later, in a corresponding passage in the *Nicomachean Ethics*, Aristotle apparently rejects homonymy of friendship:

Friendship being distinguished into these species <perfect friendship and friend-ships formed for the sake of some extrinsic good>, vulgar people will be friends because of pleasure or utility, because they are in this respect the same, whereas good people will be friends because of themselves, for <they are friends> in so far as they are good people. The good are friends, then, without qualification, whereas vulgar people are friends only coincidentally and because of their being likened to good people. (*EN* 1157ᵇ1–5; cf. 1157ª25–36)

Aristotle does not explicitly deny the homonymy of friendship here. Still, he regularly treats one form of friendship as appropriately called friend-ship primarily and properly (*prôtôs kai kuriôs*; *EN* 1157ª30–1, cf. 1156ᵇ34), while the others are so called merely because of some resem-blance (*kath' homoiotêta*; *EN* 1157ª31–2; cf. 1156ª16–17, ᵇ36–1157ª1, 1158ª19–21, ᵇ5–11).

Since mere resemblance will not satisfy the forms of account depen-dence required by CDH, it is unlikely that Aristotle will be in a position to regard the forms of friendship enumerated as core-dependent homonyms. It is striking, in any case, that in the *Nicomachean Ethics* Aristotle distances himself from the secure treatment of friendship as

---

[25] I accept a conventional dating of the *Eudemian Ethics*. See Chapter 1, n. 1.

[26] Maintaining *hêmin* with the mss. against the Revised Oxford Translation's unneces-sary emendation (Barnes 1984).

homonymous in the *Eudemian Ethics* by omitting all mention of source dependence.[27]

If Aristotle changes his mind about friendship, it will not be due to any new insight into the univocity of 'friendship'. Rather, he develops an ambivalence about whether imperfect friendships bear the sorts of relations to perfect friendship which would qualify them as core-dependent homonyms. Or perhaps, more fundamentally, he changes his mind about which relations are the right ones. It is difficult to determine exactly, because he never self-consciously articulates those relations.[28] Even so, as I shall argue, CDH provides a mechanism for making this determination.[29]

'Friendship' is in many ways like 'life'. Neither is uncontroversially non-univocal; and if shown to be non-univocal, neither qualifies as core-dependent in any obvious way. For the arguments which establish non-univocity may also tend to show non-associativity. But 'friendship' is unlike 'life' in so far as it introduces additional complexity regarding Aristotle's thinking about homonymy. His shifting attitudes about friendship may reflect a change in his thinking about the nature of friendship; or it may instead follow from a refinement in his approach to homonymy; or it may be a function of both.

The possibility that Aristotle develops or refines his account of homonymy raises some difficult issues about how his development in other areas ought to be viewed. Since he introduces homonymy in his very earliest works,[30] and even recognizes friendship as a focally connected

---

[27] In the *Eudemian Ethics*, Aristotle appeals to one of his common illustrations of associated homonymy, namely 'medical' (1236ª22–4). He makes no such appeal in the *Nicomachean Ethics*.

[28] This explains the divergence of interpretation one finds regarding the homonymy of friendship in the *Nicomachean Ethics*. Owen (1960) and Gauthier and Jolif (1970, ii. 669–86) find a continued commitment to the associated homonymy of friendship. Irwin (1987, 360) doubts that Aristotle maintains his commitment. Cooper (1977) is expressly non-committal. Still others seek a modification in Aristotle's approach to homonymy, including Fortenbaugh (1975) and Walker (1979). In different ways, both Fortenbaugh and Walker introduce a form of associated homonymy which does not involve core-dependence. Walker introduces a 'largely neglected, though no less intriguing form of homonymy' (195), which he terms 'homonymy *kath' homoiotêta*' (193). Both Walker and Fortenbaugh concede, then, that Aristotle's treatment of friendship in the *Nicomachean Ethics* marks a departure from the account of the *Eudemian Ethics*. There is logical space for Walker's non-focal associated homonymy, but resemblance cannot be its hallmark. Instances of discrete homonymy may involve resemblance without association. In my view, it is unlikely that Aristotle has merely changed his mind about friendship. For he seems to comment on a mistaken form of reasoning which would lead one to accept either univocity or core-dependence (*EN* 1155ᵇ13–15), rather than on mistaken views about the nature of friendship itself.

[29] See 4. 4 and 4. 5 below.

[30] See Chapter 1, n. 1.

core-dependent homonym very early, it is not possible to regard him as discovering or inventing homonymy and the philosophical methodologies it invites only after developing other features of his metaphysical or semantic theories.[31] This takes on special significance when we turn to consider the most difficult and disputed of all cases of core-dependent homonymy, the homonymy of 'being'.

Arguably, it is nothing other than his recognition of the homonymy of being which underwrites his introduction of a science of being qua being in *Metaphysics* iv. 4.[32] Early on, in the *Organon*, Aristotle had doubted the possibility of a general science of being (*An. Post.* 92ᵇ14, *Top.* 121ᵃ16–19, 121ᵇ7–9; cf. *Met.* 996ᵃ6, 998ᵇ9, ᵇ22, 1045ᵃ6). Because every science requires a unified subject matter (*An. Post.* 77ᵃ5–9), the homonymous character of being must not only be compatible with the sort of unity required for science but must actually provide for such unity. For mere non-univocity is insufficient. Indeed, one would normally accept non-univocity as prima-facie grounds for refusing to countenance a science in a given area. There is no general science of 'bank' having money-lending institutions and river borders in its domain. Conceivably, Aristotle's initial reservations about the possibility of a science of being derive from the thought that being is non-univocal. If it later turns out that being admits of a unified treatment so that there is after all a science of being, perhaps this will be due to its being homonymous. Here, then, instead of collapsing into simple non-univocity, homonymy preserves the unity required for science.

On this approach to the homonymy of being and the science of being qua being, Aristotle makes a discovery not about homonymy but about being. The discovery is that being is a focally connected homonym, in precisely the way that friendship was thought to be in the *Eudemian Ethics*. Indeed, the very example of focal connection used to illustrate the nature of friendship in the *Eudemian Ethics* resurfaces just after the introduction of the science of being qua being in *Metaphysics* iv. 1. Hence, it is

---

[31] It is clear that some versions of homonymy were current in the Academy before Aristotle. Speusippus evidently developed some notion of homonymy, though the evidence about its precise character is inconclusive. See Hambruch 1904; Cherniss 1944, 58–9 n. 47; Anton 1968; Barnes 1971; and Taran 1978. Plato's dialogues make sparing use of the notion of homonymy: *Phaed.* 78e1–6, *Crat.* 405d5–e2, *430b2–c1, Thaet.* 147c7–d1, *Soph.* 218b1–4, 234b5–11, *Pol.* 257d1–258a1, *Parm.* 126c6–9, *133d3–6, Phil.* 57b8–9, *Phaedr.* 265e4–266a10, *Prot.* 311b6–8, *Rep.* 333b2–4, *Tim.* 41c6–d1, 52a4–7, *Laws* 757b1–3. The notion is non-technical in most of these passages and is roughly equivalent to 'namesake'. I have marked the more technical uses with an asterisk.

[32] I discuss this introduction in 9. 2.

concluded, a realization about the homonymy of being makes possible for Aristotle a science whose possibility he had earlier denied.

There will be some plausibility to this way of thinking, on the assumptions that: (i) the science of being qua being introduced in *Metaphysics* iv is in fact incompatible with the denial of a science of being in the *Analytics*;[33] and (ii) Aristotle's attitudes towards core-dependent homonymy remain fixed and without development. The case of friendship may give us reason to doubt (ii).

2. 9 HOMONYMY'S FIRST PROMISE: CORE-DEPENDENCE AND SCIENCE

These questions about Aristotle's development point to a promise homonymy holds for Aristotle. Since he maintains that every proper science studies a class of entities realizing a single, unified universal (*An. Post.* 77ª5–9, 83ª30–5, 85ᵇ15–18, *Met.* 1059ᵇ24–7), Aristotle threatens to deny himself any opportunity for systematic or scientific study each time he assails a Platonic univocity assumption. If there is no Form of Goodness, indeed if there is no single thing common to all instances of goodness, then there will be no branch of scientific inquiry into this area. As we have seen, the same holds true of being. If there is no genus of being, there will be no science of being. Indeed, a simple, perfectly general argument seems to show that whenever univocity fails, scientific inquiry is impossible:

(A) A necessary condition of there being a science in a given domain *D* is there being a single universal realized by all the members of that domain.[34]

(B) The members of domain *D* realize a single universal only if that universal admits of a univocal account.

(C) Hence, the members of *D* will be treated by a single science only if they realize a universal capable of receiving a univocal account.

---

[33] For questions about the relation between the unity of science and Aristotle's concerns about being, see Bolton 1995.

[34] Aristotle never regards this as a sufficient condition, however. All beige things realize the universal 'beige'; but there is not science of beigeness. The demands of science include additionally genuine necessity (as opposed to accidental uniformity), explanatory priority, and basicness (*An. Post.* 73ᵇ32–74ª3). See Kung 1977 and Ackrill 1981.

If (C) is correct, then non-univocity precludes scientific investigation.[35]

As we have seen, sometimes Aristotle uses a principle in the neighbourhood of (C) to castigate the Platonists. At other times, he employs it himself, to call into question the possibility of a given science. Thus, he worries especially about whether there could be a science of all the causes, given that there is no one form of cause most basic and universal (*Met.* 995^{b}6–10, 996^{a}18–^{b}26).[36] By arguing this way, Aristotle accepts the negative results yielded by his assaults on univocity.

Still, if we think that 'life', 'being', 'goodness', and 'friendship' are non-univocal, then, if we accept (C) as stated, we also reject the possibility of scientific inquiry into them. We moreover preclude ourselves from seeking to offer unified accounts of such central philosophical notions. And we end up having only a critical role to play in philosophy.

Aristotle does not accept this result in his own philosophy. He thinks he is in a position to offer scientific accounts of life and being. So, as long as he continues to think of them as non-univocal, he must reject or modify either (A) or (B).

Core-dependent homonymy provides a mechanism for denying (A), while preserving part of its motivation. If Plato conflates several types of goodness into a single Form of Goodness, he will be guilty of an unwarranted univocity assumption.[37] None the less, we remain able to provide a scientific account of goodness if all the forms of goodness are suitably related to a core notion. Our account will not, of course, be univocal; nor will it be merely equivocal.[38] Science will be possible because there will, after all, be a single binding universal. To be sure, this universal will not bind together particulars by being instantiated by everything in the domain of the science. Instead, everything in the domain of the science

[35] Dialectic may nevertheless be possible in such cases. Still, Aristotle disparages dialectic as incapable of reaching firm, indisputable conclusions (*Top.* 155^{b}10–11). Here he evidently has weak dialectic in mind. (I accept Irwin's distinction between strong and weak dialectic, 1988, 18–21, 174–8.) It is clear that Aristotle envisages more for the study of goodness and being than the weak dialectic can deliver.

[36] In these passages, Aristotle presumes, as is common from the *Analytics*, that each distinct genus yields a distinct science; he then applies it to the four genera of causes and infers directly that no one science could take them as its object. Here Aristotle offers an especially clear instance of the inference from (A) and (B) to (C). See Ross 1924, i. 227.

[37] The strain of Plato's univocity assumption in the case of Goodness surfaces most clearly in the Analogy of the Sun (*Rep.* 507a–509c), where Plato avails himself of both moral and non-moral senses of Goodness in explaining the efficacy of the Good in generating and illuminating other objects of intellection.

[38] Leszl (1970) assumes that these are exhaustive options and is appropriately criticized for doing so by Hamlyn (1978, 1). Alston (1971) makes an analogously faulty assumption regarding analogical predication in Aquinas. Rapp (1992) also relies on an exhaustive dichotomy.

will be bound in a tight, conceptual way to the universal which forms the core of the domain. This, at any rate, is what CDH proposes.

Considered this way, homonymy promises to relax the demands on science advanced in the *Analytics* without abandoning them altogether. If he can adequately explicate and defend CDH, Aristotle will be in a position to alter the first premise of this argument precluding scientific investigation into non-univocal terms. In place of (A), he can hold:

> (A′) A necessary condition of there being a science in a given domain *D* is there being a single universal (i) realized by all the members of that domain or (ii) necessarily appealed to in a complete account of any member of that domain.

(A′) expands the scope of science without sacrificing Aristotle's commitments to commonality and priority. Although (A′) together with (B) does not entail (C), it nevertheless places sharp constraints around the subject matter of any given science. By adopting (A′), Aristotle can maintain his belief that every science is set over a single domain determined by an explanatorily basic universal. At the same time, he can reject (C) because he will not be bound to hold that every science has as its domain a group of objects realizing a single univocal universal.

Core-dependent homonymy permits Aristotle to achieve this by relaxing the qualifications on domain membership. (A′), unlike his original (A), permits a science to consider entities which do not realize a single universal, but only if their accounts necessarily make appeal to that universal. Significantly, by adopting (A′), Aristotle can continue to maintain (B) in common with those whom he criticizes for assuming univocity illicitly. Consequently, Aristotle will be able to agree with Plato that a given philosophical notion admits of scientific treatment and analysis; but the analysis will reveal complexity where Plato had seen simplicity.

2.10 HOMONYMY'S SECOND PROMISE: ORDER IN MULITIPLICITY

Aristotle's scruples about unity in science are motivated in part by the reasonable intuition that any science investigating more than one domain is bound to fail. It will fail in explanatory terms, because it will treat disparate phenomena as if they were integrated; and it will, in consequence, hover at too high a level of generality to capture the essential nature of the species it proposes to study. If, for example, there are many distinct forms of cancer which lack any common core, then the oncologist who purports to study what is common and essential to all forms of cancer is bound to be frustrated.

Aristotle's demand for unity in science is reasonable. Equally reasonable is a correlative demand for unity in philosophical analysis. If someone wishes to determine whether computers are intelligent, she will inevitably be led to reflect upon the presuppositions of her own question. She will, in short order, be forced to ask what intelligence consists in and so will need to reflect upon the question of whether there is some one universal intelligence which all and only intelligent creatures realize. If she comes away from her investigation having determined that there is not, that there are a family of related properties stitched together by forms of resemblance falling short of identity, she will come to appreciate that the texture of her initial question has changed. Now it will be inappropriately general for her to ask whether computers are intelligent. Instead, she will rightly focus her questions on whether computers are fully computational; or whether they are creative; or whether they trade in concepts; or whether they demonstrate semantic proficiency; or whether they can manipulate internal representations; or whether they have information storage capacities; or whether they develop strategies for coping with environmental stress. She may well determine that computers manifest some of these abilities but lack others; and she may now regard her original question as too general to admit of any determinate or meaningful response.

So far, then, homonymy promises to steer philosophical investigation away from unhelpful and potentially misleading generality. For it counsels awareness of multiplicity. But the role of core-dependent homonymy in science also promises a less destructive, more optimistic use in all forms of philosophical theorizing. Core-dependent homonymy promises in addition to recognize forms of connection richer than mere resemblance.

In this sense, core-dependent homonymy is usefully and importantly contrasted with a more contemporary methodology many have found attractive, namely the doctrine of family resemblance.[39] The results obtained by Wittgensteineans are often purely negative, relying on strong and compelling intuitions about non-univocity. Thus, when Wittgenstein points out that it is difficult to provide necessary and sufficient conditions for something's qualifying as a game, he provides strong prima-facie

[39] There is a temptation, in which some indulge, to treat homonymy as akin to Wittgensteinean family resemblances. Thus, Von Wright (1963, 11–12, 15–16) wrongly assimilates homonymy to family resemblance. See also Ziff 1960. A somewhat closer contemporary analogue would be Ryle's notion of a *polymorphous concept*, on which see Ryle 1951 and Urmson 1970. The notion as introduced in Ryle is suggestive, if underdeveloped; as it is developed by Urmson (in terms of 'action contents'), a polymorphous concept will be too restricted to qualify as homonymous in Aristotle's sense.

reasons for recognizing its non-univocity.[40] But it is wrong to infer, as so many of his followers have,[41] that non-univocity closes off all possibility of further inquiry. If Aristotle can establish the core-dependence of traditional philosophical concepts, then core-dependent homonymy will provide for a form of analysis possible even after non-univocity is established—indeed, a form of analysis possible only after non-univocity is established.

Homonymy's second promise, then, is methodological. If he can articulate and defend an adequate account of core-dependence in line with the demands of CDH, Aristotle will offer a way of philosophizing which is sensitive to metaphysical multiplicity but alive to the prospects of multiplicity's order.

2. 11 CONCLUSIONS

These investigations suggest that Aristotle's appeals to homonymy may be fruitful in both critical and constructive contexts. At the same time, they equally show that Aristotle's reliance on homonymy in disputed cases will be compelling only to the extent that he can answer the challenges about his methods which his critics will appropriately pose. First, he needs to show that he has a plausible method for establishing non-univocity. Second, he needs to establish that some cases of non-univocity are nevertheless associated. Third, he will need to show whether the association displays the forms of core-dependence required by CDH. Finally, when attempting to establish this association, Aristotle must guard against sliding unintentionally into a form of univocity both he and his critical targets had earlier failed to appreciate.

I begin considering the prospects for homonymy in general in the next chapter, by investigating Aristotle's most forceful technique for establishing non-univocity, namely difference in signification.

---

[40] *Philosophical Investigations*, §§69–71.
[41] See e.g. Bambrough 1961.

3

Homonymy and Signification

3.1 NON-UNIVOCITY AND SIGNIFICATION

All forms of homonymy require non-univocity. Sometimes non-univocity is immediately obvious: most, but not all, discrete homonyms exhibit their multivocity to just anyone. Many associated homonyms and some discrete homonyms are, by contrast, seductive. Moreover, the multivocity of every philosophically interesting associated homonym will rightly be disputed. If Plato thinks that goodness is univocal, he will appropriately demand from Aristotle an argument for its non-univocity. Aristotle's appeals to homonymy are justifiable only to the degree that he can provide such arguments.

In this chapter I consider Aristotle's techniques for establishing non-univocity. Because most of the homonymy indicators of *Topics* i. 15 suffice only for non-disputed contexts, it will be necessary to set them aside. I focus on Aristotle's simplest method for establishing non-univocity, namely difference in signification. Although simple in some ways, this method is also difficult and controversial, for it is not initially clear how Aristotle understands signification.

I argue first that difference in signification is sufficient for non-univocity. I argue, further, that signification is, broadly, a *meaning relation*. Consequently, difference in meaning is sufficient for non-univocity. Moreover, I maintain, difference in signification is necessary for non-univocity. Hence, difference in meaning is also necessary for non-univocity and hence necessary for homonymy.

The argument schema for this conclusion is direct and simple:

(1) '*F*' in '*a* is *F*' and '*b* is *F*' signify different things if and only if '*F*' is non-univocal.

(2) Signification is a meaning relation.

(3) Hence, '*F*' in '*a* is *F*' and '*b* is *F*' mean different things if and only if '*F*' is non-univocal.

(4) Non-univocity is necessary and sufficient for homonymy.

(5) Hence, '*F*' is homonymous in the applications '*a* is *F*' and '*b* is *F*' if and only if '*F*' in '*a* is *F*' and '*b* is *F*' mean different things.

Though simple in structure, the argument is complicated by its crucial appeal to signification as a meaning relation. This appeal is problematic, because it cannot be taken as obvious that signification (*sêmainein*) is a meaning relation at all.

Indeed, this argument schema may seem to have an immediately unacceptable consequence, one which calls into question its easy association of meaning and signification. If difference in meaning is necessary and sufficient for homonymy, then any competent speaker of a given natural language ought to be an expert arbiter in any dispute about non-univocity conducted in that language. It should then be impossible for Plato and Aristotle, both perfectly competent speakers of Greek, to dispute about the non-univocity of e.g. *dikê* (justice). Since this dispute is actual, and so altogether possible, it cannot be the case that difference in meaning is necessary and sufficient for homonymy. Nor for that matter should it be possible for Aristotle to recognize, as he plainly does, seductive homonyms; for competent speakers of a language should have authority over the meanings of the words they use in daily discourse. Hence, again, differences in meaning should not be necessary or sufficient for homonymy. But since differences in signification are necessary and sufficient for non-univocity, signification cannot be a meaning relation, as claimed.

Instead, it is sometimes urged,[1] signification should be viewed non-semantically. When he says that a word or, more commonly, a definition, signifies something, Aristotle intends a relation of *essence specification*. Since competent speakers of a language need not—and, most often, will not—be in a position to specify the essences of the general terms they employ, they will not be in a position to determine differences in signification. Hence, they will not be in a position to adjudicate disputes about non-univocity; hence, finally, they will not be able to determine whether a given term is homonymous. This is welcome, since it explains why two competent speakers of Greek, Plato and Aristotle, can disagree over whether a given term, e.g. 'goodness', is homonymous. At least one of them has mis-specified its essence.

In this chapter I address arguments for and against treating signification as a meaning relation. I begin with an investigation of Aristotle's account of signification by explicating its place in his general semantic theory. For without a clear appreciation of the general features of his

---

[1] Irwin (1982).

semantic theory, questions about whether signification is a meaning relation are aimless, floating free of any context of determination.

I then turn to questions about signification and word meaning. I argue
that signification is after all a meaning relation but that not all meaning
relations are what I will call *shallow linguistic meaning relations*, either for
Aristotle or for us. Consequently, Aristotle can introduce difference in
signification as necessary and sufficient for homonymy, without suffering
the unacceptable consequences sketched. Even so, a difficulty results from
Aristotle's recognition of *deep meaning relations*. This recognition impli-
cates him in an identification of properties and concepts which many will
find problematic. By showing how arguments intended to establish their
non-identity miss their mark, I defend Aristotle against charges that he
unacceptably conflates properties and concepts. Hence, Aristotle's practice of regarding some cases of non-univocity as simple and unproblematic is defensible; so too is his practice of engaging philosophically
disputed cases by engaging in substantive debate about the concepts
expressed by core philosophical terms.

3. 2 SEMANTIC THESES

The single most important such passage for evidence concerning
Aristotle's approach to semantic phenomena opens the *De Interpretatione*, where Aristotle draws a fairly complete analogy between words
and sentences on the one hand and types of affections (*pathêmata*) in the
soul on the other:

> First it is necessary to settle what a name is and what a verb is, and then what
> a negation, an affirmation, a statement, and a sentence are.
>
> Spoken sounds are symbols of affections in the soul, and written marks <are
> symbols> of spoken sounds. And just as letters are not the same for all, neither
> are spoken sounds the same <for all>. Nevertheless, what these are first of all
> signs of, viz. the affections of the soul, these are the same <for all>, and what these
> affections are likeness of, viz. things, these are also the same <for all>. These
> <matters> have already been discussed in the *De Anima*, for they belong to
> another subject.
>
> Just as there are, then, in the soul some thoughts which are neither true nor
> false and others which are necessarily one or the other, so also with spoken sound.
> For both truth and falsity concern combination and separation. Thus names and
> verbs themselves, e.g. 'man' or 'white' when nothing further is added, are like the
> thoughts without combination or division; for <these> are as yet neither true nor
> false. Here is an indication of this: 'goatstag' signifies something, but is not yet
> true or false, until one should add 'is' or 'is not' (either simply or with respect to
> time). (*Int.* 16ᵃ1–18)

This passage is both rich and precise in the relationships Aristotle sees between linguistic expressions, mental states, and extra-linguistic, extra-mental things.[2]

The analogy between thoughts and spoken sounds is reasonably straightforward: individual words are to assertoric sentences just as individual thoughts are to compound thoughts. The first members of these pairs are 'as yet' without truth value; in some sense they are semantically atomic. The second members are by contrast necessarily either true or false. That is, Aristotle claims that bivalence obtains for all simple assertoric sentences and their mental analogues, compound thoughts, 'either simply or with respect to time'.[3]

Partly in virtue of this analogy, this passage contains the seeds of four related Aristotelian semantic theses, each of which appears in various guises and is developed throughout the corpus: (1) *compositionality*: the semantic value of compound thoughts, as for complexes of words, viz. assertoric sentences, is a function of their sub-sentential or sub-propositional semantically relevant parts; (2) *conventionality*: the written and spoken symbols used to stand for thoughts are conventional, whereas the thoughts themselves and that for which they stand are not; (3) *relational-*

---

[2] As Kretzmann (1974, 3) points out, portions of this text 'constitute the most influential text in the history of semantics'. In view of this history, and the torrent of commentaries which constitutes it, it is difficult to say anything uncontroversial about this text, including even the claim that the *pragmata* of 16ª7 are to be construed as 'extra-linguistic, extra-mental' things. But this interpretation is assumed below. A useful compilation of an important cross section of commentaries on this passage, translated into English with critical discussion, has been provided by Arens (1984). Kretzmann (1974, 18–19 n. 6) thinks that efforts of those commentators in the Latin West before Moerbeke's translation of 1268 were seriously marred by Boethius's unfortunate rendering of both *sumbola* and *sêmeia* as *notae* (here rendered by 'symbols' and 'signs'). Although he is right that Boethius's translation could have been more faithful, Kretzmann's assessment rests in part upon his view that 'this terminological difference reflects a real difference Aristotle recognized' (5), a difference between natural signs (*semeia*) and conventional signs (*sumbola*). Irwin (1982, 25 n. 15) rightly doubts whether this distinction can be applied consistently. Cf. *Int.* 16ª17, 16ᵇ10, ᵇ20. Even so, my discussion of this passage is deeply indebted to Kretzmann's article, and has also benefited from Ackrill 1963 and Nuchelmans 1973.

[3] Aristotle's treatment of future contingents in *De Interpretatione* 9 raises the question of whether he means to restrict this general claim in some way, perhaps by excepting future-tensed assertoric sentences. Of the many approaches to this engaging chapter, relatively few have attempted to offer an interpretation sensitive to Aristotle's broader semantic theory. Among those that have, some regard Aristotle as rejecting bivalence for future contingents, and so as contradicting or implicitly restricting the general claim made here, while others have sought to offer a reading according to which he never denies bivalence and so never abandons this more general principle. The purport of *De Interpretatione* 1 would reasonably incline one to expect no rejection of bivalence in *De Interpretatione* 9; but whether Aristotle fulfils this expectation and whether, if not, he is inconsistent are matters for a separate discussion.

ism: conventional semantic units (written marks and spoken sounds) receive their semantic significance from those things of which they are symbols; finally and most important (4) *signification*: the relationship in virtue of which they receive their semantic significance consists in or involves what Aristotle calls signification (*sêmainein*).

3. 3 SIGNIFICATION

Written words are symbols of spoken sounds, which in turn are signs of affections in the soul. In virtue of these relationships, words and sentences, both written and spoken, signify. We might say that in virtue of various conventions words and sentences have meanings and referents, and we might expect Aristotle to have some analogous semantic relationships in mind in making this claim. Although this expectation will not be disappointed, it is not immediately clear what sort of semantic relationship Aristotle intends, or even whether the notions of meaning and reference are relevant to its explication.[4]

In *De Interpretatione* 1, Aristotle introduces signification only by way of shoring up his contention that no atomic semantic units are truth evaluable. He suggests that although 'goatstag' signifies something, it is not yet true or false, evidently because it is not yet a part of an assertoric sentence (*Int.* 16ᵃ16–18). Aristotle seems to imagine someone objecting to his contention that a necessary condition of being truth evaluable is being a complex semantic unit by claiming that 'goatstag' is false, on the grounds that it applies to nothing at all. His response is that although it has no referent, 'goatstag' none the less has signification and is in any case neither true nor false. If this is correct, the fact that 'goatstag' has a vacuous reference plays an essential role in the illustration; the objector might additionally have claimed, however, that 'Socrates' is true, because 'Socrates' has a referent. The rather confused underlying view seems to be that singular terms are true if and only if they have referents.[5] The force of Aristotle's response is then to point out that the objector has somehow confused reference with truth evaluability.

However that may be, in allowing that 'goatstag' has signification even though it has no referent Aristotle seems initially very close to identifying

---

[4] See Hamlyn 1977–8 for the claim that Aristotle's semantic theory does not incorporate the notions of sense and reference. See also Haller 1962 and Kneale and Kneale 1962, 45, for related discussions.

[5] Another logically possible but even odder view consistent with the text would be that non-vacuous singular terms lack truth value, while vacuous singular terms are false. But no view in antiquity entails this odd view or could even be plausibly confused with it.

signification with meaning. What 'goatstag' has cannot be reference, since part of the point of the dialectical objection is that it lacks reference, and this is a point Aristotle rightly concedes; the only other available semantic value would seem to be, then, meaning. Hence, it might be thought that not only has Aristotle disentangled meaning from reference; he also implicitly realizes that words can be meaningful even when they lack referents, and he identifies meaning with what he calls signification.[6]

This analysis not only coheres with what is said in *De Interpretatione* 1 but seems the only plausible explanation for the semantic principles Aristotle appeals to in that chapter. But Aristotle's treatment of signification does not always dovetail with the account underlying *De Interpretatione* 1, and indeed some of Aristotle's remarks about signification seem positively incompatible with that view. Thus, it would be mistaken, or at any rate premature, to identify signification with meaning or to read *De Interpretatione* 1 as providing anything like conclusive evidence that Aristotle advances a two-tiered semantic theory or even recognizes a distinction between sense and reference.

The starkest evidence of an incompatibility in Aristotle's views about signification may be represented in terms of the following problematical triad:

(1) 'Goatstag' (*tragelaphos*) has signification (*Int.* 16ᵃ16–17; *An. Post.* 92ᵇ5–7, cf. *An. Post.* 92ᵇ29–30).

(2) 'Signification' means meaning.[7]

(3) 'Manandhorse' (*himation*)[8] has no signification (*Int.* 18ᵃ19–27).

---

[6] Further evidence that signification is meaning includes: (1) Aristotle's claim that names, verbs, phrases, and sentences signify, while parts of names and verbs (*Int.* 16ᵇ28–33) and particles do not (*Poet.* 1456ᵇ38–1457ᵃ6); (2) the fact that words with the same signification also mean the same (*Top.* 103ᵃ9–10; *Soph. El.* 168ᵃ28–33; *Phys.* 185ᵇ7–9, 19–21; *Met.* 1006ᵇ25–7). Prima-facie evidence to the contrary includes Aristotle's claim that 'not-man' signifies something indefinite (*Int.* 16ᵃ29–31, ᵇ11–15), whereas 'manandhorse' signifies nothing at all.

[7] The notion of meaning invoked in this triad is intended to be quite general. *De Interpretatione* 1 suggests that it should be somehow distinguished from reference, but even this assumption is not necessary to generate this prima-facie inconsistent triad, so long as 'signification' is read univocally: 'goatstag' and 'manandhorse' seem equally to have sense but lack reference.

[8] *Himation* means 'cloak' and not 'manandhorse', but Aristotle stipulates in *De Interpretatione* 8 as elsewhere that we are to use it in a technical way. In this context he is concerned in part about whether the surface grammar of sentences reveals the number of assertions made. Within this context, he imagines a single word or phrase *x* which picks out two discrete entities, and asks how many assertions are made in the claim *x* is *F*. The entire passage runs: 'But if one name is given to two things, from which one thing does not result, there is not a single affirmation (*kataphasis*). For example, if one were to give the name "cloak" (*himation*) to a horse and man, "a cloak is white" would not be a single

If 'signification' means meaning, and both 'goatstag' and 'manandhorse' are meaningful, then (1)–(3) form an inconsistent triad. Since Aristotle explicitly advances both (1) and (3), it would seem that the interpretation embodied in (2) must be mistaken: although there is nothing which is a 'manandhorse', just as there is nothing which is a 'goatstag', the term itself is nevertheless meaningful, at any rate as meaningful as 'goatstag'. Both seem equally to be meaningful singular terms lacking referents. Thus, signification cannot be identified with meaning.

Before this is conceded, however, it is worth asking whether *any* univocal account of the semantic nature of signification is available such that: (i) it makes sense of the illustration of *De Interpretatione* 1, and (ii) is something which 'goatstag' has but 'manandhorse' lacks. In order to satisfy (i), signification cannot be reference and must have something to do with sense expression; in order to satisfy (ii), it therefore cannot be reference but evidently cannot be sense either. If so, signification can be neither sense nor reference, and so must be either minimally broader than these notions or somehow altogether divorced from them.

But even if signification is taken to be divorced from the semantic notions of sense and reference, the problem of satisfying (i) and (ii) remains, and it is not obvious that other candidates are any more successful.[9] For example, since Aristotle allows the name and formula (*logos*) to signify the same thing (*Top.* 162ᵇ37–163ᵃ1), where formula is plausibly taken as definition, it follows that names signify what definitions signify: essences (*Top.* 101ᵇ38). Thus, perhaps words signify not meanings but essences, and the signification relation is not the meaning relation, or even any related semantic notion. If things signified are distinguished from meanings or senses and regarded as essences,[10] then perhaps (1) can be shown to be consistent with (3). But given that essences are properties

affirmation. For saying this would not differ from "A horse and a man are white". So if these <the latter> signify many things and are many <affirmations>, it is clear that the first one (sc. "a cloak is white") signifies either many things or nothing—for there is nothing which is a man and a horse' (*Int.* 18ᵃ18–26). Strictly, then, Aristotle says not that *himation* lacks signification where it is taken to mean 'manandhorse', but rather that the full sentence in which it features lacks signification; even so, it is clear, given his commitment to compositionality, that Aristotle would regard the entire sentence as lacking signification precisely because its subject lacks signification.

⁹ Irwin (1982, 245–7) argues that words signify essences and that consequently signification 'cannot be meaning, because names signify essences and essences are not meanings, but belong to non-linguistic reality' (246). But the problem presented by 'goatstag' reappears for this account, but at a different level, since 'goatstag' is held to have signification, even though there is no real property to be signified by that word.

¹⁰ Irwin (1982, 246) suggests this sort of cleavage: 'Now an essence is a universal, a definable property of things in the world; it is not the sense or meaning of a name or a linguistic expression.' Irwin does not provide his reason for making this distinction, but cf. 246 n. 8.

explicitly restricted to existing entities (*An. Post.* 92^{b}4–8), such an attempted reconciliation would only introduce new problems: if signification relates words to essences, then 'goatstag' surely lacks signification.[11] Thus, (1)–(3) remain inconsistent when we divorce signification from meaning, reference, or associated semantic notions. Essences fare no better than meanings, so long as signification is read univocally in (1)–(3).[12]

This suggests either that Aristotle uses signification equivocally, or perhaps that on closer inspection 'goatstag' is meaningful whereas 'manandhorse' is not. Aristotle clearly thinks that not only words signify: the word 'man' signifies rational animal, but the entity *man* signifies rational animal as well (*Cat.* 3^{b}10–23; *Top.* 122^{b}16–17, 142^{b}27–9; *An. Post.* 85^{b}18–21; *Met.* 1017^{a}22–7, 1028^{a}10–16); clouds signify rain and smoke fire (*An. Pr.* 70^{b}10–38).[13] But such friendly equivocation is unlikely to help Aristotle in sorting out his views on the signification of 'goatstag' and 'manandhorse': both passages occur in similar contexts within the *De Interpretatione*, and both are clearly concerned with the signification of words rather than of things.

The probable resolution is that Aristotle does not after all think of 'manandhorse' as meaningful. He gives only a glimmer of an argument in explaining why 'manandhorse' fails to signify: 'for nothing,' he claims, 'is a manandhorse' (or: 'for it is not the case that something is a manandhorse', or perhaps even simply, 'for no man is a horse'; *ou gar estin tis anthrôpos hippos*). If Aristotle here means nonexistence is a sufficient condition for non-signification, he has clearly contradicted himself: part of the force of the dialectical objection in *De Interpretatione* 1 depends precisely upon the point that goatstags do not exist. Consequently, when Aristotle claims that nothing is both a man and a horse, he evidently has

---

[11] Irwin's suggestion (1982, 257–8) that it is possible to apply Aristotle's distinction between what is known to us and what is known by nature (*An. Post.* 71^{b}23–72^{a}5) to signification so that 'goatstag' signifies something to us but nothing by nature seems to me extremely fruitful. But: (i) it does not yet explain why 'goatstag' and 'manandhorse' are not equally significant to us; and, more important, (ii) it is equally available to those who think that signification is meaning, since they may hold that different senses of 'signification' are invoked in the distinct passages.

[12] Indeed, if essences are regarded as distinct from meanings, Aristotle's remarks in *De Interpretatione* 1 become difficult to interpret; consequently, signification as meaning more easily satisfies desideratum (i).

[13] Ackrill (1963, 88) has suggested that Aristotle moves sloppily between substances and words which name them by saying that both 'man' and man signify. Irwin (1982, 254–5) argues that commentators are right to find a use-mention confusion in Aristotle only if signification is meaning: 'His claim would be puzzling if we supposed that words signify their meanings; for then it would be hard to see how man could signify a meaning' (255). But (a) words could signify meanings even though substances signify essences, and (b) if meanings are in some cases essences, then there is a straightforward way in which man could signify a meaning. Hence, Aristotle need not confuse use and mention if signification is meaning.

some stronger condition in mind, for example that nothing could be both *fully* a man and *fully* a horse. He is not then introducing fantastic creatures like centaurs, but rather creatures that do not and cannot exist. The goatstag, by contrast, is the fantastic beast familiar from Aristophanes.[14] If so, the claim that 'manandhorse' has no signification entails only that self-contradictory expressions are meaningless and is fully compatible with the claim that 'goatstag' has signification, where signification approximates meaning.

The context in which Aristotle denies signification to 'manandhorse' lends further support to this interpretation. In *De Interpretatione* 8 he is concerned with the individuation of assertoric sentences, and the examples of 'manorhorse' and 'manandhorse' (both *himation*) are introduced to demonstrate that 'if one name is given to two things, from which one thing does not result, there is not a single affirmation' (*kataphasis*) (*Int.* 18ª18–19). But if 'manandhorse' were understood to apply to mythical creatures like centaurs, it is unclear why Aristotle would introduce this example as an instance of applying a name to two things 'from which one thing does not result' (*ex hôn mê estin hen*). On the contrary, if centaurs or the like were intended, we would have a case in which one thing was composed out of two halves; but these would clearly be irrelevant to the point of *De Interpretatione* 8.[15]

3.4 TWO PROBLEMS ABOUT SIGNIFICATION

These considerations provide a resolution of the inconsistent triad (1)–(3) which falls short of forcing a rejection of the prima-facie attractive suggestion that signification is closely related to meaning as opposed to reference or another, altogether non-semantic notion. But they do not establish the view that signification is meaning, and the discussion of the problems presented by 'goatstag' has brought out, importantly, that there are clear competitors to this interpretation. Most significantly, it has emerged that Aristotle, at least in some cases, thinks that words signify essences rather than meanings; and if meanings and essences are distinct, then there is an insurmountable impediment to his holding that signification is meaning, or at any rate to his holding that signification is always meaning.

---

[14] *Frogs* 890.

[15] I am doubtful about Irwin's claim, 'Both "manorhorse" and "manandhorse" have clear, definite single meanings, no less than "man" and "horse" have' (1982, 257). 'Manorhorse' does not have a single meaning, whereas 'manandhorse' does not have a clear or definite meaning.

There is first of all, then, a problem about the relationship between the prima-facie attractive thesis that signification has something to do with meaning and the thesis, equally compelling on textual grounds, that signification is signification of essences. We have some contexts, such as *De Interpretatione* 1, where signification is reasonably taken as meaning; but in other contexts Aristotle's remarks entail that words signify essences. With the further premise that essences are not meanings, we are faced with the conclusion either that (a) Aristotle uses signification equivocally or (b) the passages in which signification seems to be meaning must be reinterpreted in light of Aristotle's other commitments. But neither of these alternatives is an attractive one, and indeed the second collapses into the first.

As we have seen, *De Interpretatione* 1 is recalcitrant to any straightforward reinterpretation in terms of essences, since 'goatstag' is significant even though goatstags do not have essences. This is a consequence of which Aristotle should presumably be aware, since he elsewhere allows in a single passage that 'goatstag' has signification even though non-existent entities have no essences (*An. Post.* 92^{b}4–8). Indeed, generally, if only essences can be signified, then since non-existent entities have no essences, no word without a reference can have signification. Thus, on the view that signification is signification of essences, any significant word, precisely because it has signification, turns out to have existential import. But Aristotle claims quite self-consciously that the names of non-existent entities have signification. Consequently, and this is the first problem about signification, signification seems to be a relation which both has and does not have existential import. Hence, if signification is signification *only* of essences, then 'signification' is used equivocally by Aristotle. Therefore, if meanings are divorced from essences, 'signification' does not admit of a univocal account in Aristotle.

There is, of course, no reason in principle why Aristotle should not use 'signification' equivocally, even if non-semantical considerations of essentialism are not introduced. We use 'meaning' equivocally: words have meanings; people mean things when they use words; the evidence means that such-and-such is the case; getting to the airport on time means taking the early train; funnel clouds mean tornadoes threaten; and neighbourhood bullies mean trouble. Even so, within this list, some senses of 'mean' are core, and others are derived or analogical, and reductions between various members may be forthcoming. Further, given the technical role 'signification' seems to play within Aristotle's semantic theory, it would be surprising if he were to use it rather cavalierly in similar contexts, as if it were univocal.

But a situation in which Aristotle equivocates on 'signification' results only from the assumption that meanings and essences have nothing to do with one another, or more precisely that no essences can be identified with meanings; for it is only on this assumption that the claim that signification is meaning appears incompatible with the claim that essences are signified. Further investigations into Aristotle's semantic theory may render this assumption dubious; and it is in any case necessary to inquire whether this assumption is reasonable, as well as whether there is evidence for supposing that Aristotle accepts or rejects it.

A second problem about signification also invites deeper investigation into Aristotle's semantic theory. If signification is at least in some instances signification of essences, we can expect signification to be a relation not graspable in every instance by competent speakers of the language. Not all competent speakers are aware of the essences of the natural kinds to which they refer, so some may have positively false views about the essential natures of things associated with the words they use. When a scientist engages in an inquiry to establish the essence of kind *F*, she is in part—on the view that signification is signification of essence—attempting to determine what it is that a word signifies, because she is trying in part to determine first what all *F*s have in common, and second which of the properties that *F*s share are explanatorily fundamental.[16] But this seems to run afoul of the conventionality thesis introduced in *De Interpretatione* 1 and further explicated in the second chapter of the same work: 'A name is a spoken sound significant (*sêmantikê*) by convention, without time, none of whose parts is significant when separated . . . by convention because among names none is significant by nature, but only when it has become a symbol' (*Int.* 16ᵃ19–27, omitting 16ᵃ21–6). The claim that names are significant by convention does not seem immediately compatible with the claim that real essences—real properties discovered only by scientific inquiry—are significates.[17] Conventions governing signifier use seem to proceed quite independently of such investigation.

That said, one cannot infer directly from the conventionality requirement that signification is not signification of essences. In *De*

---

[16] Aristotle does not believe that all necessary properties are essential, because he argues that the *proprium* is necessary, even though it is not part of the essence (*Top.* 102ᵃ18–24, 128ᵇ34–6). Cf. Irwin 1988, 507 nn. 38, 39. All humans are necessarily capable of grammar, but being capable of grammar is not part of the human essence. This suggests that in addition to being necessary, essential properties must be causally and explanatorily basic in the sense that they are prior to other necessary properties.

[17] Bolton (1985) argues initially that the conventionality requirement entails the falsity of Irwin's (1982) view that signification is signification of essences. Although I do not think this inference is justified, I think the conventionality requirement shows that there may be more connection between the traditional view of signification as meaning and the heterodox view of signification as signification of essences than has been supposed.

Interpretatione 2, Aristotle contrasts significance by convention (*kata sun-thêkên*; 16ª19, 26–7) with significance by nature (*phusei*; 16ª27), thus making clear that his target in claiming that names are significant by convention is the sort of view found, for example, in Plato's *Cratylus*, according to which there is some natural connection, like onomatopoeia, between names and the things they name. Since signification of essence need not be signification by nature, conventionality need not be regarded as incompatible with that view.

It might be countered, however, that the conventionality requirement is incompatible with the signification of essences at a deeper level. The conventionality requirement of the *De Interpretatione* seems to require that competent speakers of a language know what the names in their language signify precisely because they implicitly grasp the conventions regarding the uses of those names.[18] But if all, or even some, names signify real essences, then ordinary speakers must grasp those essences in grasping the conventions governing the applications of those names. Since it is scientists, and not ordinary speakers, who discover essences, this result is surely unacceptable.

None the less, when he says that names are significant by convention, Aristotle does not claim that the relevant conventions are those employed by ordinary speakers. So, on the condition that Aristotle is willing to jettison the assumption that ordinary speakers understand the deep signification of all the words they use, he will be in a position to affirm both that names are significant by convention and that signification is signification of essence.

Does Aristotle meet this condition? Some passages suggest that he does

---

[18] Bolton (1985) suggests that Aristotle accepts the following three claims: (1) some names signify real essences; (2) competent speakers, as such, know what *all* of the names which they use signify by convention; (3) competent speakers do not know, as such or in general, the natures of the real essences which their names signify. He argues that these three propositions are compatible only if 'competent speakers as such do somehow know *which* real essences are *by convention* signified by those names which do signify real essences but do not necessarily know what the fundamental *natures* of those real essences are' (154). Aristotle accepts (1) and (3); but it is unclear whether he holds (2), and the evidence Bolton adduces on its behalf is less than convincing. He claims, '*Every* name, for Aristotle, is significant due to an agreement or convention among its users as to what it signifies. So speakers can be expected to know, at least implicitly, what *all* of the names which they use signify' (158). But *De Interpretatione* 1 and 2 say simply that names are significant by convention, not that the conventions established by the language community at large fix the signification of the names they use. Hence, conventionality as such is too weak to justify the claim that a competent speaker of a language will understand the significance of every term she uses. We arrive at that conclusion only via the ancillary assumption that the conventions which fix signification are the conventions governing meaning; but that is the question at issue.

not (*Top.* 103ᵃ9–10; *Soph. El.* 168ᵃ28–33; *Met.* 1006ᵇ25–7). First, Aristotle's compositionality thesis rightly includes the proviso that only the *semantically relevant* portions of words contribute to the meanings of complex expressions. In this context, he claims that word-parts, even word-parts which if used independently would be meaningful, do not signify (*Int.* 16ᵇ28–33). This suggests that Aristotle means to distinguish meaningful from non-meaningful sounds by appealing to signification, and so that he aligns signification and meaning. Second, words said to have the same signification regularly have the same meaning (*Top.* 103ᵃ9–10; *Soph. El.* 168ᵃ28–33; *Met.* 1006ᵇ25–7); this seems to be a further non-accidental correlation between signification and meaning. Finally, Aristotle is uncomfortable with the suggestion that co-referential terms should for that reason alone be regarded as signifying the same (*Met.* 1054ᵃ13–19). Hence, again, although there is no formal contradiction inherent in Aristotle's conception of signification, the notion seems overtaxed: Aristotle evidently expects signification to play one role in explaining meaning and a further non-related role in scientific inquiry.

This double-duty, then, suggests a second problem about signification. Conventionality may be taken narrowly or broadly, so that the relevant conventions governing signification are either those established by scientists in a position to discern essences or are those implicitly grasped by ordinary speakers. If taken narrowly, the result seems to be that ordinary speakers do not know what the terms they use signify; if taken broadly, the result seems to be that signification cannot be signification of essence. Aristotle does not seem comfortable with either of these results.

Like the first problem, the second problem recommends an investigation into Aristotle's conception of the relationship between essences and meanings. Both problems rest in part upon the initially plausible assumption that meanings are distinct from essences, even for those natural kind terms whose referents Aristotle clearly holds to have essential natures. But the plausibility of this assumption may be diminished by a sufficiently complex analysis of the meanings of such terms in Aristotle's semantic theory.

This investigation follows and is conducted within the framework *De Interpretatione* 1 establishes. Given Aristotle's commitment to relationalism, it is first of all necessary to determine what the *relata* of the signification relation are. If the signification relation is the meaning relation, then given Aristotle's conventionalism, we should expect these *relata* in some cases to be conventional marks or sounds on the one hand, and their meanings or intensions on the other. But given Aristotle's analogy between the semantic evaluability of such marks or sounds and the *pathê-mata* of the soul, we can equally expect this relation to obtain between mental states and those meanings or intensions. If the signification is signification of essences, then we should expect these *relata* to include written marks or sounds on the one hand and language-independent essences on the other.

Moreover, despite the apparent cleavage between these two views, it is clear that the claim that the signification relation is a meaning relation is wholly compatible with the claim that words signify essences, on the condition that essences turn out to be intensions and that the contents of assertoric sentences featuring kind terms are partially constituted by the essences of those natural kinds.

Now, assertoric sentences about natural kinds do not exhaust the contentful sentences there are. Thus, Aristotle may simultaneously claim that signification is signification of essences and that signification is meaning, and he may do so without holding in addition that *all* meanings are essences. One expects Aristotle's analysis of thoughts to be embedded in a broader analysis of content; to the extent that this analysis shows that Aristotelian contents include essences, we will be justified in concluding that signification is meaning even though, in some cases, signification is signification of essences.

3.5 HOMONYMY AND WORD MEANING: THE CASE AGAINST

Aristotle's approach to signification invites further reflection on the question of whether difference in meaning is relevant to establishing non-univocity. We have seen that Aristotle sometimes *calls* words homonymous and elsewhere speaks as if words were homonyms. It is therefore natural to talk about the many senses of a homonymous term;[19] in hundreds of contexts where Aristotle appeals to homonymy, translations into various modern languages do just this by recasting his arguments in terms of sense or meaning. This tendency, however natural, may seem unfortunate when we reflect on Aristotle's actual accounts of homonymy, which uniformly treat things and not words as homonyms. Although, as we also have seen,[20] there is a deflationary way of understanding words as homonyms and a correlatively harmless way of understanding his arguments as appealing to senses or meanings, from this perspective all such treatments must be regarded as relaxed extensions of the strict doctrine.

---

[19] Some commentators who regard homonymy as pertaining to sense or word meaning include Hintikka (1971; 1973, ch. 1), Owen (1960 and, more guardedly, 1965*a*); and Barnes (1971), who makes difference in meaning a defining feature of homonymy.

[20] See 2.1 above.

This working compromise is in some ways unhappy, in so far as it suggests that strictly speaking homonymy is not a semantic phenomenon at all, that *F*s are associatively homonymous *F*s just because they equally realize the property *F*-ness; they are discretely homonymous *F*s if there is no one genuine property *F*-ness they realize. If *F*s are associatively or discretely homonymous to the degree that they are (or are not) members of the same functionally determined kind *F*, then their homonymy may be thought to have nothing at all to do with word meaning. Appealing to homonymous words as displaying distinct or related senses will then be systematically misleading, even perniciously so: this will mask the fact that Aristotle's doctrine is a doctrine about kinds and the properties which bind them, and not at all about words and their meanings.

For these sorts of reasons, some have argued that homonymy is not intended to mark different senses of words at all.[21] The case here can be made more precise, and more challenging, as follows. Since homonymy occurs when we have distinct definitions of *F*s (either associated or discrete), it is determined by the essences of kinds: homonymous *F*s have distinct essences. To use an example of a discrete homonym, when variously applied 'crane' picks out entities of altogether different genera and so with different essences. Since its extension covers distinct essences, 'crane' is a homonym. Further, essences are not meanings, but rather real properties in the world.[22] Therefore, since it is a function of distinctness in essence, homonymy does not register distinctions in meaning or sense.

---

[21] Irwin 1981, 524.

[22] Irwin (1982) understands signification as non-semantic for two related reasons. In addition to the argument discussed in the text, his principal arguments are these:

(A) Signification and Definiteness

1. Negative names and verbs (e.g. not-man and not-walking) are indefinite names and verbs (*Int.* 16ª29–33, ᵇ11–15).
2. Indefinite names and verbs do not signify one thing; they signify 'something in a way one and indefinite' (*Int.* 19ᵇ9).
3. Hence, negative names and verbs signify something indefinite.
4. Negative names and verbs do not mean anything indefinite; they mean something perfectly definite.
5. Signification is identical with meaning only if meaning relations are identical with signification relations.
6. If (4), then meaning relations are not identical with signification relations.
7. Hence, signification is not meaning.

(B) Signification and Uniqueness

1. 'Manorhorse' and 'manandhorse' do not signify some one unique thing. ('Manorhorse' signifies two things, while 'manandhorse' signifies nothing at all.)
2. 'Manorhorse is white' has just one meaning.
3. 'Manandhorse is white' is perfectly meaningful; it too has one meaning.
4. Signification is identical with meaning only if meaning relations are identical with signification relations. *(cont.)*

The argument so construed has a strong conclusion: that homonymy is not about word meaning at all. I will argue that Aristotle's account of homonymy does not warrant this strong conclusion. If successful, my counter-argument shows that homonymy may be intended to mark distinctions in meaning or sense; but it is a further question as to whether it actually does. Some passages beyond those already canvassed seem to have just this import. Accordingly, I will turn to them after mounting my counter-argument.

The strong argument has several discernible steps:

(1) The name and account (*logos*) signify the same thing. (Thus, 'pity' and an account of pity, φ, signify the same thing.)

(2) An account expresses the definition corresponding to the name.

(3) Hence, the name and definition signify the same thing.

(4) Definitions signify essences (*Top.* 101ᵇ38).

(5) An essence is a universal, an explanatorily basic property common to all the members of a kind.

(6) A meaning is not a universal, an explanatorily basic property common to all the members of a kind.

(7) Signification is identical with meaning only if meaning relations are identical with signification relations.

(8) If (6), then meaning relations are not identical with signification relations.

(9) Hence, signification is not meaning.

Proposition (1) is true, as is (2). Consequently, (3) follows immediately, on this assumption. Step (4) is equally unproblematic—as long as we realize that Aristotle accepts various forms of definition. Propositions (5) and (6), then, are the crucial premises: if it can be shown that no essences are meanings, then at least some homonyms will not concern sense or meaning. I will investigate (6) especially in order to show why this argument falters; I will then turn to forms of signification to show how proponents of

---

5. If (2) and (3), then meaning relations are not identical with signification relations.
6. Hence, signification is not meaning.

In the text, I provide reasons for denying (A 4) and querying the inference from (A 2) to (A 3); I also provide reasons for denying (B 3).

a sense-based account of homonymy can accommodate the reasonable remarks made in support of this premise.

Aristotelian accounts paradigmatically state essences by placing the thing defined into its genus: 'it is necessary for the one defining well to define through the genus and differentiae' (*dei men dia tou genous kai tôn diaphorôn horizesthai ton kalôs horizomenon*; *Top.* 141ᵇ25–7; cf. 142ᵇ22–9, 143ᵃ29–ᵇ10). Since nothing is in more than one genus, Aristotle infers that no two definitions of the same thing are possible (*Top.* 141ᵇ22–142ᵃ2, 144ᵃ11–15, 151ᵃ32–ᵇ2, ᵇ15–17, 152ᵃ23–ᵇ5). Aristotle displays the connection between essence, definition, and genus quite clearly in *Topics* vi. 4, when suggesting techniques for determining whether one who has offered a definition 'has stated the essence' (*eirêke to ti ên einai*; 141ᵇ24–5):

First <one must examine> whether he has not rendered the definition through what is prior and better known. Since the definition is given in order to know what has been said, and we know not through what happens <to be the case> but through what is prior and better known, just as in demonstrations (for all teaching and learning occur in this way), it is clear that the one not defining through these sorts of things has not defined. But if one has not <provided a definition in terms of what is prior and better known>, there will be more than one definition of the same thing. For it is clear that the one defining through what is prior and better known has defined better, so that both of these will be definitions of the same thing. This sort of thing, however, does not seem <to be so>; for there is one essence (*to einai hoper estin*) of each of the things that are. So if there are more definitions than one, the essences specified as corresponding to each of the definitions for the things defined will be the same. But these are not the same, since the definitions are different. It is clear, then, that the one not defining through what is prior and better known has not defined. (141ᵃ27–ᵇ2)

The argument proceeds in part by assuming both that each definition will display an essence and that two definitions are the same if, and only if, they specify the same essence.

This passage presents some difficulties,[23] but its main purport is clear. Aristotle assumes a uniqueness condition, such that definitions must specify the unique essence of the *definiendum*. Since every definition purports to state an essence, and since two definitions are the same if they specify the same essence, more than one definition of the same thing is impossible. Hence, Aristotle concludes, rather strongly, that the one who has not specified has not defined at all (*Top.* 141ᵃ31, ᵇ1–2).

This conclusion is rather strong, since it holds definition to a standard

---

[23] See Poste 1866 *ad loc.* for a discussion.

which Aristotle elsewhere relaxes or rejects altogether. Indeed, in this very passage Aristotle allows that some definitions may be better (*beltion*) than others and that those which state the essence in terms of what is prior and better known are precisely the ones which are better (*Top.* 141ᵃ32–4). For some pedagogical purposes, better definitions are worse definitions. Essence-stating definitions framed in terms of what is prior and better known are better *simpliciter* since they are more scientific (*epistê-monikôteron*; *Top.* 141ᵇ15–17). 'Nevertheless', Aristotle concedes, 'it may be necessary to render an account (*logos*) for those incapable of knowing through these sorts of ways through what is known to them' (*Top.* 141ᵇ17–19), where the accounts clearly qualify as definitions,[24] even though one must not overlook that 'those who define in this way cannot specify the essence of the thing defined' (*Top.* 141ᵇ23–4).

Definitions which fail to specify the essence are inferior definitions; but they are nevertheless definitions.[25] (Inferior knives are knives all the same; perhaps inferior knives are better for children.) If so, then when Aristotle says that those who fail to state the essence have failed to provide a definition, he means only that they have failed to provide the best sort of definition, or the most secure definition, or the most scientific kind of definition. His acknowledgement of better and worse definitions shows that Aristotle views (4) as potentially misleading: definitions do state essences, though not every definition states an essence. Consequently, even granting for the moment the rest of the argument, the case against understanding homonymy in terms of meaning is significantly curtailed. On the correct reading of (4), we would be entitled to infer (9) only if it had a form of generality it lacks. For on the assumption that not all definitions state essences, the fact that homonymy marks definitional differences will not require homonymy to register only differences pertaining to the essences of homonyms. Hence, even granting (4), we do not end up with an uncontroversial argument for (9).

It is worth noting in this connection that Aristotle also explicates sameness of definition partly in terms of sameness of signification, where this may indicate an interest in the more recognizably semantic phenomena of synonymy and meaning. Aristotle holds, for example, that in framing definitions, it is necessary to replace the word being defined in a truth-preserving way. Definitions must replace the word being defined, since otherwise we would have uninformative, circular definitions (*Top.*

---

[24] The phrase *tôn toioutôn horismôn* at *Top.* 141ᵇ19 clearly refers back to the accounts (*logoi*) at 141ᵇ19.

[25] Perhaps it is best to think of *horos* itself as an associated homonym: all definitions reveal something about the thing defined; core definitions reveal the essence.

147^{a}29–^{b}25). When the word being defined is replaced, the terms employed in the *definiens* must have the same force[26] and must signify the same thing (*An. Post.* 92^{b}26–34). Taken together, these passages suggest a fairly liberal, non-technical attitude towards both definition and signification on Aristotle's part.

3. 6 BETTER KNOWN TO US/BETTER KNOWN BY NATURE

When he appeals to different forms of definitions and signfication, Aristotle relies upon a general framework of concept possession outlined in the *Posterior Analytics*. Aristotle often claims that in scientific reasoning we begin with what is known to us and proceed to what is better known by nature:

The road <toward principles and causes> is naturally from what is better known and clearer to us (*hêmin*) to what is clearer and better known by nature (*phusei*); for what is known to us is not the same as what is known *simpliciter*. Consequently, it is necessary for us to progress in this way from what is less clear by nature, but clearer to us, to what is clearer and better known by nature. (*Phys.* 184^{a}16–21, cf. *Phys.* 188^{b}32, 189^{a}5, 193^{a}5; *EN* 1095^{a}2–4, 1095^{b}2–3, 1098^{b}3–8; *Top.* 105^{a}16, 141^{b}4, 141^{b}25; *An.* 402^{a}3–10, 413^{a}11–16; *An. Post.* 71^{b}9–16, 71^{b}32–72^{a}5)

In dialectical and scientific contexts alike, our natural starting point will be the beliefs we have; but our conclusions need not agree with these beliefs, so that what was initially clearer and better known to us may not finally be clear or known to us, or even believed by us at all. What is clearer and better known 'by nature' will be what is known to us as well when we have grasped principles and causes, that is, when we have come to understand the way things are independent of our initial beliefs.

In speaking this way, Aristotle allows in a general way that our beliefs about e.g. dogs may evolve: our initial beliefs, based on sense perception, may incorporate false views about the natures of dogs which are corrected

---

[26] 'It is necessary to substitute words which are able to do the same thing (*ha to auto dunatai*), words for words, words for phrases and a word for a phrase, and always to take a word for a phrase; for thus is the setting forth of the definitions easier. For example, if it makes no difference to say that what is supposed is not of the same genus as what is believed or that what it is to be what is supposed is not <what it is to be> what is believed (for the same thing is signified), one must give what is supposed and what is believed rather than the phrase' (*An. Pr.* 49^{b}2–9).

through the course of our inquiry into the nature of dogs. Hence, our initial beliefs about dogs encode information about how dogs appear to us, even though these appearances are themselves subject to revision.[27]

In drawing a distinction between how things appear to us and how they are by nature, Aristotle suggests that some contents incorporate *modes of appearance*. When S thinks that dogs are carnivorous, the content of her thought incorporates the form of dogs and the property of being carnivorous; but the form of dogs appears to S a certain way, and a complete representation of the content of her thought should take notice of this fact. This mode of appearance may typically be perceptual, although it need not be, and may be represented:

$$\{\text{doghood}^{MA}, \text{being carnivorous}^{MA}\};$$

and the complete belief relation:

$$R(S, \{\text{doghood}^{MA}, \text{being carnivorous}^{MA}\}).$$

This commitment complicates Aristotle's analysis, but for good reason.

First, Aristotle thinks that forms are identical with essences. Essences are for Aristotle those properties which are not only necessary to a particular kind but are also fundamental in the sense of explaining the existence of other properties invariably realized by members of that kind (the *propria*). But he also thinks that forms are constituents of the contents of thought. If so, then any thought about the natural kind 'dogs' involves an awareness of the essence. Hence, S thinks something about dogs only if she knows what the essence of dogs is; it would seem to follow, then, that since anyone can have a thought about dogs, nearly everyone must be familiar with their essence. If this were so, scientific investigation would be pointless, because the results of any given inquiry would be known in advance by those posing the questions.

This unfortunate consequence is blocked by an introduction of modes of appearance, by a way of understanding something as it is known first to us, but not as it is known by nature. It may be true that when S thinks something about dogs, the form of dogs enters into the content of her thought. But since S can grasp this form under some accidental concomitant embodied in the mode of presentation, it does not follow that she has

---

[27] It is sometimes thought that Aristotle's methodological commitments constrain him to embrace at least those appearances which are universally shared. Nussbaum (1982 and 1986) and Evans (1987) both commit Aristotle to this methodological principle, and both rely in part on *EN* 1172ᵇ36–1173ᵃ1, where Aristotle says, 'That which seems to all [sc. to be good], this we affirm' (in Evans's rendering, 'What seems to everyone to be the case, we declare to be true'; *ha gar pasi dokei, taut' einai phamen*). But Aristotle does not intend any sort of general methodological principle in this passage.

grasped the essence of what it is to be a dog in having this thought. On the contrary, *S* will have grasped the essence only when the manner in which this form is known to her coincides with how it is known 'by nature', that is, how it is in fact independent of any beliefs about it. This will occur, however, only at the end of a scientific inquiry.

Aristotle's distinction between things as they are known to us and as they are known by nature, together with its attendant introduction of modes of appearance, is reflected in a distinction he draws in the *Posterior Analytics* among several types of definitions:[28]

Since a definition is said to be an account of what something is, it is clear that one sort <of definition> will be an account of what a name or other nominal (*onomatôdês*) account signifies, for example what 'triangle' signifies. When we comprehend that it is, we inquire why it is; but it is difficult to grasp this if we do not know that it is. The cause of this difficulty has already been given, viz. we do not know whether it is or not, except coincidentally. (An account is one in either of two ways: by being conjoined, like the *Iliad*, or by making one thing clear of one thing, not coincidentally.)

Thus one definition of definition is the one stated, but another sort of definition is an account which makes clear why something is. Hence, the former signifies but does not prove, and it is clear that the latter will be like a demonstration of what something is, differing by position from demonstration. For saying why it thunders differs from saying that there is thunder; in one case one says, 'because fire is extinguished in the clouds'. What is thunder? The noise of fire being extinguished in the clouds. Hence the same account is said in a different way, and in this way it is a continuous demonstration, and in this way a definition. Again, a definition of thunder is noise in the clouds, and this is the conclusion of a demonstration of what it is. But the definition of immediates is an indemonstrable positing of what something is.

There is then, one definition as an indemonstrable account of what something is; one is a deduction of what something is, differing in aspect from demonstration; and a third is the conclusion of a demonstration of what something is. (*An. Post.* 93ᵇ29–94ᵃ14)

Different definitions reflect different degrees of scientific awareness, and distinct contents correspond to these distinct levels of awareness.

Aristotle claims first that one sort of definition states what a name or other nominal account signifies, and he contrasts this with the sorts of definitions involved in demonstrations. Part of this contrast consists in Aristotle's suggestion that one can know what one sort of definition a name or nominal expression signifies without knowing whether the thing

---

[28] *Posterior Analytics* 93ᵇ28–94ᵃ10 seems to list four distinct types of definitions but is followed by a summary, 94ᵃ11–14, which lists only three.

(ostensibly) named exists. Answering to this epistemic level will be one form of definition, a form which consequently makes no existential claim whatsoever. Presumably goatstags, as well as entities which do exist and do have essences, admit of such definitions. But not only non-existent entities can be so defined: Aristotle's own example is a triangle.

When he carves out this first class of definitions, Aristotle therefore extends his claim that signification need not involve awareness of the essence. In addition to those cases where there are no essences to be signified, words picking out entities with essences signify, even though those essences do not figure into the definitions which are accounts 'of what a name or other nominal (*onomatôdês*) account signifies' (*An. Post.* 93ᵇ30–1). Aristotle's introduction of this form of definition may seem initially incompatible with his claim that forms constitute contents. If forms are essences, and thought involves forms as contents, then thought seems immediately to involve essences. But the form of signification to which Aristotle appeals permits him to deny that thought involves awareness of the essence while maintaining simultaneously that forms figure into contents. When *S*'s thought that Pinkie is a dog is not scientifically informed, the form of doghood can figure into her content under a certain mode of presentation, a mode which is an accidental concomitant of that form. In endorsing a form of definition which neither has existential import[29] nor displays the essence of the thing defined, Aristotle implicitly allows definitions of entities under modes of presentations. Such definitions correspond to his epistemic distinction between things better known to us and things better known by nature, and invite further reservation about treating signification mono-dimensionally.

---

[29] Bolton (1976) denies this claim and argues that Aristotle's view of natural kind terms bears some salient similarities to the direct reference theories of Kripke and Putnam, in so far as they all maintain that the meaning of a natural kind term is fixed by a quasi-indexical introduction of the referent. He suggests, however, that Aristotle differs from these theorists in supposing that the mechanism by which reference is initially determined involves stereotypical characteristics (as opposed to e.g. initial baptism), which nevertheless are 'built into the account of the reference of the term . . . in such a way that their possession, individually or collectively or disjunctively, is neither necessary nor sufficient for belonging to the reference of the term' (543–4 nn. 52–4). Aristotle seems to reject any claim that understanding the meaning of a term entails knowledge that it has referents. Bolton further argues that there are no nominal definitions of entities which do not exist, e.g. goatstags; this would follow from Aristotle's holding that a definition is always of what something is, while there are no accounts of non-existents. But in allowing a form of definition of what a word signifies, Aristotle allows nominal definitions of things which do not exist. There are no real definitions of such entities. Even so, goatstags and the like fall into the same category of entities which have essences but whose definitions do not mention them. Such definitions, recognized at *Topics* 141ᵇ22–142ᵃ9, are available equally to entities with and without essences.

An implicit reliance on modes of appearance also surfaces in Aristotle's conception of the role of imagination (*phantasia*) in perception and action.[30] If *S* is a butterfly collector, and mistakes a glimmer of light on an oddly coloured leaf quivering in the wind for a rare butterfly, and thus sets off to capture it, she will have made an error. She will have had a false belief. Aristotle's teleological account of action makes this simple analysis plausible. *S* has a particular goal in mind and, responding to a particular sensory stimulus, acts to fulfil that goal; without her stimulus having a certain appearance to her, the action in which she engages seems inexplicable. As a butterfly collector, *S* does not engage in pursuit of flickers of light but has a general desire, statable as a major premise of a syllogism which rationally reconstructs her action (e.g. rare butterflies are to be captured), coupled with a particular belief that what is present to her provides a means of fulfilling that desire (there is a rare butterfly) (*EN* 1147ᵃ24–ᵇ5; *An.* 434ᵃ16–21). Her action is intelligible, and appropriately described by us, only if the environmental stimulus has the appropriate appearance for her.

Aristotle recognizes that, as the example illustrates, imagination need not be correct (*An.* 428ᵃ18, 428ᵇ1–4). Indeed, that it need not be correct on all occasions seems essential to its role in explaining action. Actions motivated by misperceptions, hallucinations, or delusions become explicable when the role of imagination in motivating action is taken into account. Imagination's role in these contexts seems to involve the attribution of description-theoretic, rather than *de re*, thoughts to agents. Indeed, it involves representing entities in certain ways, on the basis of memories and perceptual stereotypes, with the result that we apprehend them under certain modes of appearance. Here too Aristotle seems to recognize a problem for treating all definitions and forms of signification as equivalent.

3. 7 CONCEPTS AND PROPERTIES

If this is correct, the conclusion (9) that signification is not meaning will be in some ways at odds with Aristotle's approach to signification and definition. Even if these worries were waived, however, we would have difficulty accepting the crucial premise (6), the claim that meaning is not a universal, an explanatorily basic property common to all the members

---

[30] Cf. Schiffer (1987, 175), who objects to 'the cavalier way in which the "Fido"–Fido theory ignores the connection between semantics and psychology'.

of a kind. For (6) seems indifferent to the implicit appeals to modes of appearance which Aristotle makes both in his account of definition and in his philosophy of action.

Certainly (6) seems attractive if we are willing to accept a sharp distinction between concepts and properties. For if meanings are concepts and concepts are not properties, then we have reason for distinguishing between them. Still, not all philosophers of language accept a sharp dichotomy between meanings and real properties.[31] Others endorse (6) and diagnose some serious errors in philosophy as deriving from a failure to recognize its truth.

It is not immediately clear why we should regard Aristotle as among those who endorse (6); and it is hard to identify any obvious mistake in accepting (6), unless it is supposed to be that a commitment to (6) requires one to believe that a priori conceptual analysis will sometimes lead to discoveries about synthetic property identifications.[32] But one who endorses (6) need not accept this. On the contrary, someone could endorse (6) without making any such assumption.[33]

This is important for our understanding of Aristotle's attitude towards (6). Because he does not offer self-conscious methodological principles about the nature of the analysis he practises, it is necessary to examine the kinds of definitions he offers. Some of his definitions quite naturally make reference to physical processes or properties not to be discerned except

[31] See e.g. Carnap 1956 and Bealer 1982.

[32] This seems to be what Putnam (1981, 207) believes: '[Philosophers have] conflated properties and concepts. There is a notion of property in which the fact that two concepts are different (say "temperature" and "mean molecular kinetic energy") does not at all settle the question whether the corresponding properties are different. (And discovering how many fundamental physical properties there are is not discovering something about concepts, but something about the world.) The concept "good" may not be synonymous with any physical concept . . ., but it does not follow that being good is not the same as being P, for some suitable physicalistic (or, better, functionalistic) P. In general, an ostensively learned term or property (e.g., "has high temperature") is not synonymous with a theoretical definition of that property; it takes empirical and theoretical research, not linguistic analysis, to find out what temperature is (and, some philosopher might suggest, what goodness is), not just reflection on meaning.' Of course, it would be foolish to believe that one could settle matters of empirically discernible identities by appealing to the lexical meanings of terms as they have developed in natural language. It would also be unwarranted to saddle Aristotle with this kind of foolishness: he never reasons in this way and never gives any indication that he would respond favourably to such a misguided approach.

[33] If the property 'water' is identical with the property 'being H_2O', then no amount of conceptual analysis could reveal this fact. But this will have nothing to do with any question of property and concept identification. Indeed, it cannot, unless one of the properties is also a concept, for the synthetic identity is held to obtain between properties, not a property and a concept. See Strawson 1987 on property and concept identification. His view, although heterodox, is to be preferred. Strawson's criticisms of Wiggins (1984) on this score are wholly convincing.

through empirical research.[34] Others definitions make no such reference.[35] Aristotle characteristically insists that the type of definition required is partly a function of a particular sphere of inquiry. Thus, in the *De Anima* he notes:

> The physicist and dialectician will define each of these <that is, each affection of the soul> differently. For example, the second <would define> anger as the desire for retaliation, or something of this sort, whereas the first <would define it as> the boiling of the blood or heat around the heart. One of these specifies the matter, and the other the form and account (*logos*). For the account is the form of the thing; but if the form is to be, it must be in matter of a certain sort. (*An.* 403ᵃ29–ᵇ3)

There are, then, dialectical and physical definitions. Aristotle complicates matters further by proceeding to categorize the province of the physicist, mathematician, and first philosopher in terms of whether they treat attributes separated from matter, without clearly allying the dialectician with any one of them (*An.* 403ᵇ9–16).[36] The kinds of definitions appropriate to these various inquiries do not uniformly make reference to material realizations, even while it is acknowledged that the forms do not exist unless realized.

Aristotle's willingness to countenance different types of definitions in different spheres suggests a first approach to his attitude towards (6). Clearly he resists the impulse to suppose that all definition can proceed by mere linguistic analysis. He further requires explicit reference to physical properties for some types of definition. He therefore would not want to permit a physicist qua physicist to suppose that she could expose the nature of heat merely by reflecting on the linguistic meaning of heat. Yet this much circumspection does not take us all the way to (6). The physicist's definitions might not be of concern to a philosopher, for two related reasons. First, the sphere of inquiry need not be one philosophers investigate. Second, and more important, where there is overlap, the philosopher may offer definitions uninformed by conditions of material realization, if the properties in question can exist separate from matter (*Phys.* 194ᵃ12–ᵇ9, 198ᵃ22–31; *Met.* 1037ᵃ14). Hence, Aristotle's awareness of physical definitions does not provide the sort of evidence needed to regard him as committed to (6), at least not in the way some contemporary philosophers have wanted to be. Moreover, he need not commit the

---

[34] In the *Posterior Analytics*: thunder at 93ᵇ12–14, lunar eclipses at 90ᵃ15, 93ᵃ30–3, 93ᵇ6–7; in the *De Anima*: colour at 419ᵃ9–11, anger at 403ᵃ25–ᵇ9.

[35] I doubt, for example, that Aristotle hopes for empirical definitions of goodness, substance, actuality, form, or even soul.

[36] Hicks (1907, 200) supposes that the mathematician is a 'mere dialectician'. This presumes that in this passage Aristotle has in mind only a weak concept of dialectic.

mistakes associated with a rejection of (6), since nothing about rejecting it as such would enjoin him to presume that linguistic analysis narrowly construed is sufficient for revealing the essences of terms of concern to philosophers.

Instead, Aristotle can easily distinguish, as we often do, between *deep* and *shallow* meaning. Shallow meaning is, roughly, what a competent speaker of a natural language requires for her competence. Deep meaning, by contrast, is the meaning revealed by investigation—whether conceptual analysis or empirical study—which permits us to grasp the essence. This distinction permits us to say, for example, that although Euthyphro pontificates about piety, he does not even know what it means. In speaking this way, we do not accuse him of being an inept speaker; we accuse him, instead, of being shallow.

A distinction between deep and shallow meaning thus turns crucially on whether a term has been investigated sufficiently to reveal its essence. Of course, someone may be a perfectly competent speaker of a language without being able to offer an analysis on demand. On the empirical side, one will understand the meaning of 'water' without being able to point out that water is essentially H_2O, if indeed that constitutes its essence.[37] Similarly, one may well know the meaning of 'mercy' adequate to most purposes, even while being utterly incapable of specifying the necessary and sufficient conditions of something's counting as merciful. Unless we stipulatively restrict the notion of meaning to what is required for linguistic competence, it will be perfectly appropriate to regard an inquiry into the nature of mercy as an inquiry into the meaning—the deep meaning—of 'mercy'.[38] Thus, importantly, we may in some cases expect an account to display a deep meaning, while in other more relaxed contexts we are, and should be, content with shallow meanings.

This distinction between shallow and deep meaning is not intended to be a technical one. It is, however, intended to display how Aristotle can comfortably use 'signification' in stronger and weaker senses without fear of contradiction or confusion. In some cases, signification requires essence specification of a sort which requires conceptual analysis or empirical research. In other cases, in comparatively relaxed contexts,

---

[37] I doubt that we should think of water's essence this way. I am sympathetic to LaPorte (1996).

[38] Thus, when philosophers write articles on e.g. 'The Meaning of "Meaning"', they are concerned not with lexical meaning, with capturing in a descriptive way what most speakers of English intend to convey when they use the word 'meaning'; rather, they have some interest in the essence of meaning, whether e.g. it is exhausted by the internal intentions and other mental states of the subject who grasps it. This notion of meaning thus outstrips shallow meaning; it is the deep or analysed meaning of 'meaning'.

signification may require only an ability to indicate the lexical meaning of a term. In both cases, Aristotle may be understood as treating signification as a meaning relation.

3. 8 CONCLUSIONS: FORMS OF SIGNIFICATION, FORMS OF MEANING

Aristotle's co-ordinated distinctions between (i) what is better known to us and what is better known by nature and (ii) definitions which signify essences and those which do not show that he recognizes levels of signification which march in step with our notions of meaning. We think that some forms of meaning are deep and that some are shallow. It is, consequently, no argument against Aristotle's regarding signification as a meaning relation that he sometimes treats signification as essence specification. For meaning relations are sufficiently elastic that they accommodate essence specification as a form of deep meaning.

This conclusion holds true on the assumption that it is possible to regard concepts as identical with the properties which are the essences specified as deep meanings. Aristotle nowhere gives an indication that he rejects such an identification, and he often proceeds as if he regarded it as unproblematic. Moreover, contrary to the arguments of some contemporary philosophers, Aristotle is free to accept this identification without embarrassment. For investigation moves us from shallow meanings, from things signified as they appear to us, to deep meanings, to things signified as they are by nature.

This explains why competent speakers can disagree about homonymy. Competent speakers can share shallow meanings without agreeing about what analysis may reveal regarding deep meanings. It equally explains why competent speakers can nevertheless agree about simple cases of homonymy. In some cases, shallow meaning suffices: certainly all of the non-seductive discrete homonyms Aristotle mentions are obvious to any competent speaker. No one thinks a crane at a construction site is a kind of bird; and no one thinks that a professor is sharper than a shard of pottery. In many cases, shallow meaning suffices for establishing non-univocity; in interesting cases it never does. Interesting cases require investigation and analysis.[39] For this reason, disputes about homonymy

---

[39] Aristotle makes this explicit at *Metaphysics* 1004ᵇ1–4. At 1004ᵇ3–4 he points out that it belongs to the philosopher alone to determine in how many ways something is spoken, that is, to determine cases of homonymy. Since he cannot have trite cases in mind, Aristotle evidently intends the philosopher to be an investigator (*episkepsomenos*) who determines whether a given disputed or controversial philosophical term counts as homonymous.

are only as easy to resolve as conceptual analysis and empirical investigation are to perform. These activities are difficult. Still, Aristotle and his
opponents equally presume that they can be successful.

Aristotle is consequently justified in assuming that difference in signification is sufficient for establishing homonymy. For difference in
signification is sufficient for non-univocity, and non-univocity is sufficient
for homonymy. Still, if the arguments of this chapter are correct,
difference in signification may be a complex affair. In particular, analysis
may uncover differences in deep meaning which are not registered in
ordinary linguistic practice; and appeals to difference in shallow meaning
may give way to univocities uncovered by analysis and investigation.
Hence, appeals to difference in signification may be disputed in any number of ways, by those who hold out for second-order univocities and by
those misled by superficial sameness of lexical meaning.

More important, difference in signification can be used only to establish non-univocity. Consequently, appeals to difference in signification
cannot be used to establish association or core-dependence. Thus, even
though Aristotle rightly relies on difference in meaning to establish
homonymy, he cannot immediately infer that meaning relations will be
relevant to establishing association or core-dependence. I turn now to a
consideration of the general framework of core-dependent homonymy in
Aristotle.

4

Core-Dependent Homonymy

4. 1 UNITY IN MULTIPLICITY

Among homonyms, those which are associated hold the greatest philosophical interest. If he can establish both non-univocity and core-
dependence for some central philosophical concepts, Aristotle will
justifiably claim to have introduced a powerful methodology for rejecting
Platonism without adopting a purely negative or destructive attitude
towards philosophical analysis.

Thus far we have seen that Aristotle has a framework for establishing
non-univocity. We have not yet determined whether this framework has
the broad application Aristotle hopes for it; nor have we even shown that
Aristotle is right to claim that a single philosophical concept of any
importance is non-univocal. This is because we cannot know whether a
given concept is non-univocal before performing an analysis of it; and
Aristotle's appeals to difference in signification do not by themselves provide such an analysis. Instead, they presuppose that the accounts, once
provided, will reveal difference of signification. So his framework for
establishing non-univocity is a framework, and nothing more.

If we assume that Aristotle will in some cases be able to establish non-
univocity, we will have attained a primarily negative result unless we can
also show that he can in those same cases establish association and core-
dependence.[1] That is, unless Aristotle can establish that some homonyms
associate, homonymy will lack many of the constructive applications
which Aristotle presumes it has.

In this chapter, I introduce a framework for establishing core-
dependence by articulating and defending Aristotle's method for finding
unity in multiplicity. I do so in an effort to answer the problems for association and core-dependence we have identified.[2] In particular, in this

---

[1] It is incorrect to say that Aristotle's results would be *wholly* negative. For in order to
establish non-univocity for at least some range of cases, Aristotle will have to produce positive accounts which reflect genuine difference in signification.

[2] See 2. 6 for a discussion of the initial problems of core-dependent homonymy.

chapter we seek a satisfactory specification of CDH. Recall that CDH is a purely formal approach to core-dependent homonymy such that:

> *a* and *b* are homonymously *F* in a core-dependent way iff: (i) *a* is *F*;
> (ii) *b* is *F*; and either (iii a) the account of *F* in '*b* is *F*' necessarily
> makes reference to the account of *F* in '*a* is *F*' in an asymmetrical
> way, or (iii b) there is some *c* such that the accounts of *F*-ness in '*a* is
> *F*' and '*b* is *F*' necessarily make reference to the account of *F*-ness in
> '*c* is *F*' in an asymmetrical way.

A satisfactory account will specify the modality to which CDH appeals, while respecting non-univocity by avoiding the introduction of second-order univocity.

In explicating Aristotle's approach to these matters, I develop a general framework for core-dependent homonymy such that: (i) relations of association are determinate but open-ended, and (ii) relations of core-dependence are not so close that Aristotle unintentionally accepts a sophisticated, second-order form of univocity when establishing association. In this chapter, I assume what is logically prior: that Aristotle can first of all establish non-univocity for some range of philosophical concepts.[3]

In subsequent chapters I use the account developed here to appraise the force of Aristotle's appeals to homonymy and multivocity in polemical contexts. I stress that in elucidating Aristotle's general framework, I remain neutral about whether any given philosophical application succeeds or fails.

---

[3] It is hard to determine how the account I develop relates to the alternatives sketched in Leszl (1970). In his clearest statement of the possible interpretations of 'focal meaning', Leszl (1970, 123–6) mentions (i) 'the synonymy account', according to which a term in various uses 'designates at bottom just one nature'; (ii) 'the homonymy account', according to which each paraphrase of a distinct use of a term 'designates a single nature or essence which is different from the nature shown by any other of such paraphrases'; (iii) 'the complexity account', according to which only the primary use designates a single nature, 'while in its secondary uses a new conceptual content is added to the original designation'; and (iv) 'the mixture account', according to which focal connection 'is something between homonymy and synonymy (taken in their wider sense), i.e. something between equivocity and univocity, and this either because it is neither of them or because it is a sort of combination of both'. One problem is that these alternatives are not, as stated, all mutually exclusive; another problem is that they are not exhaustive; a final problem is that they provide a general framework without specifying the substantive elements of core-dependence. I find elements of both Leszl's second and fourth alternatives attractive.

4. 2 CORE-DEPENDENT HOMONYMY: A FIRST APPROXIMATION

Whatever its independent merits, Aristotle's functional determination thesis provides a way of understanding the theoretical framework Aristotle relies upon when distinguishing discrete, non-accidental homonyms. To the extent that this thesis is defensible, Aristotle is justified in claiming cases of homonymy which are neither associated nor immediately obvious. Similarly, and more generally, whenever he can establish difference in signification, Aristotle can justifiably claim non-univocity. And wherever he establishes non-univocity, he also establishes some form of homonymy. The chore, then, is to show that among all homonyms, some display systematic connection of philosophical interest.

We have noted that when he illustrates the homonymy or multivocity of key philosophical concepts like being, Aristotle frequently employs the homonymy of 'health' or 'healthy' as his key illustration. He must, then, think of these as particularly clear and uncontroversial cases of core-dependent homonymy. They are indeed relatively uncontroversial, at least in so far as acknowledging them as cases of core-dependent homonymy has no immediate anti-Platonic import. Their clarity and usefulness as illustrations of a broader philosophical method prove more difficult to establish.

It is, however, paramount that we appreciate at least how Aristotle intends cases such as health to illustrate associated homonymy in general. For Aristotle appeals crucially to the homonymy of health in a number of central passages, including most notably in *Metaphysics* iv. 2 when arguing for the multivocity of being:

Just as everything which is healthy is related to health (*pros hugieian*), some by preserving health, some by producing health, others by being indicative of health, and others by being receptive of health; and as the medical is relative to the medical craft (*pros iatrikên*), for some things are called medical because they possess the medical craft, others because they are well-constituted relative to it, and others by being the function of the medical art—and we shall also discover other things said in ways similar to these—so too is being said in many ways, but always relative to some one source (*pros mian archên*). (*Met.* 1003ª34–ᵇ6)

Unlike 'body', 'health' is a core-dependent homonym, since its various occurrences coalesce around a core notion.

The accounts of 'healthy' as it crops up in various contexts should reflect this coalescence. If we provide accounts of 'healthy' in 'healthy practice', 'healthy complexion', 'healthy glow', 'healthy regimen', 'healthy portion', 'healthy salary', 'healthy relationship', and 'healthy frame of mind', we will need to relate them all to some one principle. There must therefore be a base sense or base case of being healthy which is in some-how prior to all of these derivations. Perhaps 'healthy person' serves as such a case. If so, we may say that healthy complexions and healthy glows are indicative of the health of a healthy person; healthy practices and regi-mens are sometimes productive of health and at other times preserve it; healthy bodies are not corrupt and so are capable of being healthy, and are thus 'receptive of health'.

Core-dependent homonyms are non-discrete homonyms which display a theoretically significant form of association. So we know that core-dependence must incorporate at least this much:

> Core-Dependent Homonymy (CDH$_1$): x and y are homonymously F in a core-dependent way iff: (i) they have their name in common, and (ii) their definitions do not completely overlap, but (iii) they have something definitional in common.

The example of health provides slightly more content to the notion of definitional overlap appealed to in clause (iii). Hence:

> CDH$_2$: x and y are homonymously in a core-dependent way F iff: (i) they have their name in common, (ii) their definitions do not com-pletely overlap, and (iii) there is a single source to which they are related.[4]

CDH$_2$ is an improvement over CDH$_1$, but not much of one. CDH$_2$ (iii) requires a relation to a single source (*mia archê*) and so provides some indication of the nature of definitional overlap to which CDH$_1$ appeals. Yet Aristotle clearly permits a range of distinct relations among hom-onymous terms, as his account of health illustrates.

---

[4] Aquinas usefully distinguishes between two ways of understanding association (*SCG* i. 34; cf. *De Veritate* 2. 11 *ad* 6 and 21. 4 *ad* 2, and *ST* i. 13. 5). (He consistently and help-fully treats discrete homonymy as equivocity and associated homonymy as what he calls analogical predication. His account of analogical predication is not equivalent to analogy in Aristotle. See Chapter 1, n. 3 on analogy in Aristotle.) The two forms of association may be termed source-dependent analogy and ordered analogy:

A. Source-Dependent Analogy

SDA: a and b are analogically F in a source-dependent way iff: (i) a is F; (ii) b is F; and (iii) there is some c such that the accounts of F-ness in 'a is F' and 'b is F' necessarily make reference to the account of F-ness in 'c is F' in an asymmetrical way.

B. Ordered Analogy

OA: a and b are analogically F in an ordered way iff: (i) a is F; (ii) b is F; and (iii) the account of F in 'b is F' necessarily makes reference to the account of F in 'a is F' in an asymmetrical way.

SDA, but not OA, permits association between non-core instances. Typically Aristotle blurs this distinction by considering only cases in which some core instance is mentioned.

Aristotle never characterizes in an abstract way the nature of the relations homonymous terms must bear to the core notion around which they revolve; nor does he provide a principle for determining which relations are sufficiently strong to establish genuine association; nor indeed does he specify what makes one notion core with respect to the others. His non-performance here opens him to several critical questions.

First, how are we to determine when terms genuinely satisfy CDH and when they do not? For instance, 'healthy' in 'healthy regimen' and 'healthy diets' is homonymous. What about 'healthy' in 'healthy salary' and 'healthy appetite'? Perhaps someone with a healthy appetite consistently eats in ways injurious to his long-term health. So perhaps these are really discrete and not core-dependent homonyms. Or perhaps we can simply *construct* relations which each bears to a core notion. We might say, for example, that a healthy salary is the sort of salary which provides the material conditions for pursuing a healthy life. Then 'healthy' in 'healthy salary' will stand in the relation of 'providing the material conditions for permitting one to engage in healthy activity' and so will be, in an indirect sort of way, productive of health. Or we might say, pursuing a more narrowly semantic connection, that a healthy salary is a robust salary and that robustness implicitly appeals to the notions of flourishing evident in a core notion of human health. So we say that 'healthy' in 'healthy salary' stands in a semantic relation, perhaps attenuated, to the core notion of health. As for 'healthy appetite', we may say that a healthy appetite is one which in some circumstances, for example in a growing child, is beneficial to the realization of human health. Then we have identified some relations 'healthy' bears to a core notion of health, and will be entitled to conclude that the occurrences in question qualify as core-dependent homonyms.

The problem should be obvious: this approach to core-dependent homonymy is profligate. That we can construct some dummy relation or other is not sufficient for core-dependent homonymy. Paradigmatically discrete homonyms will trivially stand in some relation or other to a 'core' notion; but they are not therefore core-dependent homonyms. (Everything stands in some relation or other to everything else.) CDH_2 (iii) does not yet tell us how to avoid allowing junk relations to suffice for establishing core-dependent homonymy.

Nor does CDH_2 (iii) specify what constitutes being a core notion, or what Aristotle calls the 'one source' (*mia archê*). Aristotle's examples presume an asymmetry that is somewhat difficult to capture. A core case of being *F* should be core minimally in the sense that non-core cases must make reference to the core case in their account, while the core case need not make reference to the non-core cases in its account. So an account of 'healthy appearance' will in some sense include the account of 'healthy' in

'healthy person'. It is unclear why an account of 'healthy person' is pro-
hibited from making reference to some of the features which go into determining that someone has a healthy appearance, e.g. their colour, stature,
or robustness. Aristotle needs a conception of priority to underwrite the
wanted asymmetry. CDH_2 (iii) does not by itself provide this notion of
priority.

The precise contours of Aristotle's notion of core-dependent homonymy therefore turns on how CDH_2 (iii) unfolds. I will develop my analysis of Aristotle's account in two stages. First, I will discuss Aristotle's
criteria for distinguishing bogus from genuine relations of association.
Second, I will explicate a kind of priority suitable for capturing
definitional primacy. I will proceed by considering extensions and explications of CDH_2 (iii) with an eye towards developing a plausible general
formulation of CDH.

4. 3 RELATIONS OF ASSOCIATION: ADEQUACY CONSTRAINTS

As we have seen, it is not difficult in general to provide a formal
specification of Aristotle's conception of CDH_2 (iii).[5] F_is and F_{ii}s will be
core-dependent homonyms only if (i) the definition of F_i includes an
appeal to F_{ii}-ness; (ii) the definition of F_{ii} includes an appeal to F_i-ness; or
(iii) there are some third Fs, F_{iii}s, and the definitions of both F_i and F_{ii}
make appeal to F_{iii}-ness. Thus, healthy complexions and healthy regimens
will be core-dependent because definitions of 'healthy' in these cases will
equally make appeal to some core notion of being healthy, as being indicative and productive of what is healthy.

This general formal account heads in the right direction. Even so, it
does not explain what it is for one definition to make appeal to another.
Indeed, the problem of profligacy re-emerges here, since we have no con-
trols on the sort of appeals definitions are permitted to make to other
definitions. If savings banks were always as a matter of fact located within
five hundred miles of river banks, someone might mistake this as essential
to them and offer an account according to which savings banks were those
institutions located within five hundred miles of river banks where money
is kept and traded. This would have the result of turning clearly discrete
homonyms into core-dependent homonyms. Of course, Aristotle does not
want this result. The point in the present context is not that he is saddled

---

[5] This formal account proceeds along the lines sketched by Irwin (1981) and Fine
(1988).

with it, but rather that the formal account of associativity thus far says much too little. On the formal account, river banks and savings banks may count as core-dependent homonyms since the definition of savings bank makes mention of river banks. Yet Aristotle surely holds that there are restrictions concerning what definitions require and do not require.[6]

What strictures on genuine definition provide the right result for core-dependent homonymy? To begin, core-dependent homonyms manifest relations absent between discrete homonyms. CDH_2 (iii) suggests that these relations will be somehow definitional. Some clue to the sorts of relations Aristotle envisages should therefore emerge from a consideration of the sorts of definitional connections whose lack renders discrete homonyms discrete. Especially useful in this respect are the discrete, non-accidental homonyms recognized by Aristotle in certain philosophical contexts, since their discreteness requires special pleading.

Aristotle's defence of the discreteness of discrete, non-accidental homonyms is derived from FD, his functional analysis of kind membership and individuation. As noted, FD not only determines which entities are of distinct kinds but also groups functionally overlapping entities into the same kinds. Hence, one can conclude in a negative way that core-dependent homonyms cannot be members of the same functionally specified class. For then we would have univocity, and so synonymy.

The same holds true of signification. 'Healthy' in

(1) Socrates is healthy.

(2) Socrates' complexion is healthy.

and

(3) Socrates' regimen is healthy.

cannot signify the same meaning. For if they did, we would have univocity and so not homonymy. So we can conclude immediately that the forms of association can be neither 'having the same function as' nor 'signifying the same as'. For these introduce forms of unity, rather than forms of unity in multiplicity.[7]

What is wanted, then, is a specification of some asymmetrical relation

---

[6] See e.g. *Physics* i. 2, 185^{b}2–3: 'The account of infinity appeals to (*proschrêtai*) quantity, but not substance or quality.'

[7] For these reasons, I do not agree with Rapp (1992, 537–8): 'Die Pros-hen-Relation an sich setzt demnach weder ein Derivationssystem noch sonst eine Hierarchisierung von Seinsberichen voraus. Vor allem bietet Aristoteles mit dieser auf der Ausrichtung auf einen beispielhaften Fall beruhenden Ähnlichkeitsbeziehung eine zum Gattungsgefüge alternative Form von Einheit an.'

R, such that, necessarily, all non-core instances of a homonym bear R to the core:

> Necessarily, if (i) a is F and b is F, (ii) F-ness is associatively homonymous in these applications, and (iii) a is a core instance of F-ness, then b's being F stands in R to a's being F.

The positive constraints on R are, then, that it be asymmetrical, that it be open-ended in the sense of admitting new instances of non-core homonyms, and that the accounts of all non-core instances appeal to the account of the core notion. And, of course, R must respect the negative constraint of avoiding an unwanted introduction of univocity.

4. 4 CAJETAN'S PROPOSAL

Certainly one control on R should concern ways in which one account (*logos*) can depend upon another. That is, R must capture the explanatory priority of the account of a core instance of a homonym relative to the accounts of all derived instances. Perhaps recognizing this connection to priority in explanation, Cardinal Cajetan offered a simple, defensible specification of R in terms of the Aristotelian four-causal scheme, a scheme which is centrally concerned with conditions of explanatory adequacy:[8]

Analogy of this type [analogy of attribution, or *pros hen* homonymy][9] can come about in four ways, according to the four genera of causes (calling for the moment the exemplary cause the formal cause). It may occur with respect to some one denomination and attribution that many things stand in different ways to one end, to one efficient cause, to one exemplar and to one subject, as is clear from the examples of Aristotle's *Metaphysics* iv. For the example of health in *Metaphysics* iv refers to the final cause, while the example of medical in the same place refers to the efficient cause, the analogy of being, also in the same place, to the material cause, while the analogy of good, introduced in *Ethics* i. 7, refers to the exemplary cause. (*De Nominum Analogia* 2.9)

---

[8] Citations of the four causes are intended to answer complementary why-questions. See *Physics* 198^{a}14–21.

[9] Cajetan follows the customary medieval practice of treating homonymy as an instance of analogy. Though this is not Aristotle's practice (see Chapter 1, n. 3), Cajetan's meaning is clear enough. Cajetan accepts a traditional division of analogy into: (i) analogy of inequality; (ii) analogy of attribution; and (iii) analogy of proportionality. His attempt to understand the analogy of attribution in terms of the four causes appears to be an innovation, as Ashworth suggests (1995, 64). For a critical treatment of Cajetan, see McInerny 1992.

While Cajetan's illustrations seem badly garbled, the root idea, I shall argue, is exactly right. Moreover, the procedure for arriving at a specification R is also promising. Cajetan rightly begins by surveying the examples of core-dependent homonymy in an effort to specify R in the appropriately unified way.

Cajetan's illustrations are unfortunate because they treat each individual example of homonymy as somehow constrained to just one of the four causal relations. Thus, he thinks that all derived cases of 'medical' somehow refer to the efficient cause. The types of relations mentioned by Aristotle, however, do not bear this out. For example, a scalpel (*machairion*) counts as 'medical' not because it is related by an efficient cause to medicine, but because its function is given by the role it plays in medical practice (*Met.* 1003^{b}1–3, 1061^{a}3–5). Something else, a medical doctor, for example, will qualify as medical because she possesses the craft in question (*Met.* 1003^{b}1–2). So one should not suppose that all individual instances of core-dependent homonyms are related by some one kind of cause to the core notion. Instead, Aristotle supposes that different non-core instances of the same core-dependent homonym can bear different causal relations to the same core.

Extracting what is correct from Cajetan's proposal, while ignoring its idiosyncrasies, we have the following fully general specification of R, which I shall call *four-causal core primacy*:

> FCCP: Necessarily, if (i) a is F and b is F, (ii) F-ness is associatively homonymous in these applications, and (iii) a is a core instance of F-ness, then b's being F stands in one of the four causal relations to a's being F.

I defend this proposal in three stages. First, I show how it develops naturally out of some of the illustrations of core-dependent homonymy Aristotle provides. Second, I consider apparent problems with others among those same illustrations, only to argue that contrary to initial appearances, the problem cases actually strengthen FCCP. Finally, I show how FCCP satisfies the general constraints on an adequate specification of R we have established, by arguing that it is both suitably definite and appropriately open-ended.

Taking up Cajetan's suggestion, we may extrapolate from the two principal examples of core-dependent homonymy introduced by Aristotle. Some things are called healthy because they preserve health (*tô(i) phulattein*), others because they produce health (*tô(i) poiein*), others by being indicative of it (*tô(i) sêmeion*), and still others because they are receptive of it (*dektikon autês*) (*Met.* 1003^{a}35–^{b}1). Similarly, some things are called medical because they possess the craft (*tô(i) echein*), others because they are naturally suited to it (*tô(i) euphues einai pros autên*),[10] and still others because they are a function belonging to it (*tô(i) ergon*) (*Met.* 1003^{b}1–3). These examples specify *R* variously. Certainly, no first order natural relation straddles the full range of them.

Even so, there is a consistent theme through all of Aristotle's examples: all stand in an appropriate four-causal relation with the core notion. In some cases this is obvious. Thus, an account of medical in:

> (4) The scalpel is medical.

will appeal to the core notion of medical craft, so that it may be rewritten as:

> (5) The scalpel is a knife whose function is to cut in medical procedures.

Account (5) makes explicit that a scalpel counts as a scalpel precisely because its function (*ergon*) is medical. This appeal to function is ineliminable in the sense that nothing counts as a scalpel which lacks it. In this case, the core-dependent homonymy is evident because the final cause of a scalpel is easy to specify relative to the core notion of medicine. This case, then, conforms to FCCP and so provides confirming evidence of it.

Much the same can be seen in the cases of diets and regimens. Thus, the account of 'healthy' in:

> (6) Socrates' regimen is healthy.

will make appeal to a core notion of health, because of its being productive of health. It will therefore be rewritten as:

> (7) Socrates' regimen is productive of health.

Its being productive of health is, of course, its standing in an efficient causal relation to health. This case too, then, helps confirm FCCP.

Other cases are not so obvious. A complexion is indicative (*sêmeion*) of health, and for this reason it counts as healthy; perhaps vitamins are

---

[10] Aristotle's claim here is somewhat obscure. He probably means (i) that some people count as medical because, even though they are not doctors in the sense of having the appropriate craft training, they nevertheless have a knack for healing (thus, those with the appropriate abilities who are versed in home remedies count as medical, even if they are not proper doctors), or, more likely, (ii) someone who has the right form of intelligence together with other appropriate dispositions will count as medical even prior to training (sometimes adults refer to children as 'scientists' while faculty members call certain undergraduates 'professors' or 'senators' in much this sense). Cf. *An.* 421^{a}21–6; *PA* 666^{a}14; *GA* 748^{b}8; *Poet.* 1455^{a}32; *Pol.* 1327^{a}33; *Rhet.* 1390^{b}28; cf. also Plato, *Rep.* 409e3 (also a connection with medicine, and probably quite similar to Aristotle's meaning) and 455b–c.

healthy because they preserve or guard (*phulattein*) health. Neither of these relations—'being indicative of' or 'guarding'—is obviously an instance of one of the four causal relations. They therefore seem at odds with FCCP and so challenge Cajetan's proposal.

These cases should nevertheless be understood in accordance with FCCP. For FCCP does not require that the core of every homonym be the causal source of its non-core instances. That is, although it is in some sense an *archê*, the core notion need not be causally prior to the derived cases. FCCP requires only that the derived instances stand in one of the four causal relations to the core notion, without specifying in addition the direction of the causality. This should already be clear from the notion of health as applied to regimens. A regimen produces health in the sense of being an efficient cause. So while Socrates' regimen is not itself efficiently caused by Socrates' health, Socrates' regimen and Socrates' health nevertheless stand in an efficient causal relation.

The case will be nearly the same with guarding. Something's guarding health, first of all, contributes to the health of an organism by maintaining it. (A healthy diet preserves good health by contributing nutrition required to maintain proper functioning of bodily systems.) So, in this sense, guarding counts as an instance of efficient causation in the same way that a regimen does. Further, because not all efficient causation needs to be understood as sufficient for the production of a given result, guarding can count as efficient causation without its bringing about a change from illness to health in a patient. Thus, for example, Aristotle allows one's decision to find someone in the marketplace to be an efficient cause of finding that person; he thinks this decision explains the occurrence, if the agent's goal is reached (*Phys.* 193[b]36–197[a]5, 199[b]24–5). Yet no decision to find or meet anyone is by itself sufficient for its own success. Hence, the event, if it occurs, is efficiently caused by the decision without being necessitated by it alone.[11] In the same way, guarding health contributes centrally to the health of an individual. Consequently, by being this form of salient cause, it qualifies as an efficient cause. It is, therefore, no counter-example to FCCP. On the contrary, it provides additional confirming evidence.

Similar remarks can be made about something's 'being indicative of', although, in this case, the causal efficacy heads from the core outward.

---

[11] I argue here only that Aristotle allows that co-causes, or salient causes, count as efficient without being sufficient. So far, then, it is still possible for him to insist that complete efficient causes, fully specified in all particulars, are sufficient for their effects. The point I make is not that no efficient cause is sufficient, but that not all are. For another sort of attack on the sufficiency of efficient causation, see Sorabji 1980, ch. 2.

That is, Socrates' robust complexion qualifies as healthy because it is indicative of a healthy constitution. Of course, it is not accidental that this is so: healthy complexions are caused by healthy organisms. Still, as we have already seen, we do not want to hold that a healthy constitution is by itself sufficient for a healthy complexion. Other factors can defeat a causal process which occurs for the most part. Nor would we want to say that a healthy constitution is necessary for a healthy complexion. For some signs are misleading. The kinds of causal connections Aristotle permits between efficient causes and effects are, however, compatible with non-uniformity. For this reason, it is appropriate to understand the relation in question as an instance of efficient causation. Hence, this illustration, too, supports rather than undermines FCCP.

So far we have seen examples of efficient and final causal relations. Examples of material causal relations are less obvious.[12] Perhaps, though, this is why Aristotle allows that something receptive (*dektikon*) of health qualifies as healthy in a derived sense. Although he does not explicate what he has in mind, typically Aristotle characterizes a material substrate as receptive of a form when thinking of that substrate as a material cause (*Met.* 1018^{a}23–32, 1023^{a}12–14 (an especially clear instance), 1056^{a}25–7). In this sense, muscle is called healthy by being the material cause of a healthy organism. Similarly, blood will be called healthy by being a material cause of that same organism. At the same time, nothing prevents blood from being healthy in the sense of being indicative of health. So, first, some derived homonyms are material causes, and, second, some derived homonyms are doubly derived by standing in more than one of the four causal relations to the core instance. Hence, Aristotle's illustrations suggest that derived homonyms may stand in at least three of Aristotle's four causal relations: final, efficient, and material.

None of Aristotle's illustrations is an obvious instance of a formal causal relation. The question therefore arises as to whether he regards it as inappropriate for a formal causal relation to serve as a basis for core-dependent homonymy. In some sense, it clearly cannot: if a and b are both F, and the accounts of F-ness in both applications of the predicate are the same, then a and b are synonymously, and not homonymously, F. On the other hand, if a and b are both F, and the accounts of F-ness in both applications of the predicate are not the same, then a and b are homonymously, and not synonymously, F.[13] But in that case, we do not have formal causation; for, it may be argued, x's being F is a formal cause of y's being F only

---

[12] Material causal relations present a special difficulty, since Aristotle offers non-equivalent glosses on the nature of material causation. See Graham 1987, 156–8.
[13] See I. 1.

if *x* and *y* are synonymously *F*. If I threaten a clever but indolent professor with a sharp knife, then the knife's being sharp does not formally cause the professor's being sharp, even if the threat causes the professor to shed his indolence in favour of keenness. Conversely, if Socrates causes me to become healthy by his good example, then perhaps his being healthy is a formal cause of my being healthy; but then we are synonymously healthy.

This is Aristotle's point when he insists that homonymy alone is not an impediment to unity in science. Lack of unity results, rather, from 'the accounts (*logoi*) [of a homonymous term] being neither one nor referred back to one [central core]' (*Met.* 1004ᵃ24–5). So a sufficient condition for lack of unity in science is non-associated, or discrete homonymy. Conversely, if there is a single science of *F*-ness, then there is either some one account (*logos*) for all the *F*s in the domain, or all the *F*s in the domain associate around a single core. From this it can be inferred that, where the domain in general admits of scientific investigation,[14] synonymy is sufficient for science, core-dependent homonymy is sufficient for scientific investigation, and discrete homonymy is sufficient for lack of scientific unity. Since sameness of account is sufficient for synonymy, perhaps we should conclude that formal causation is altogether incompatible with homonymy.

This conclusion would restrict FCCP by precluding the possibility that a non-core homonym could ever stand in a formal causal relation to its core instance. For it treats formal causation as sufficient for synonymy, and synonymy and homonymy of all kinds are mutually exclusive. In one sense, this consequence may be welcome. For FCCP specifies *R*, where that relation should be understood as something short of formal identity. Indeed, although we have not introduced it as a formal constraint on *R*, one would expect *R* to be grounded in a so-called extrinsic denomination, that is, one in which a subject is called *F* not because it realizes *F*-ness in an intrinsic way, but because it stands in some suitable relation to *F*-ness.[15] (The author of a journal article may be intelligent in an intrinsic way; if we care to refer to the article itself as intelligent, we do so only by way of an extrinsic denomination.) If FCCP admitted formal causal relations, then we would permit *R* to be grounded in intrinsic denominations. Intrinsic denominations open the door to synonymy and so seem inappropriate as ways of grounding *R*.

---

[14] Not every collection of entities admits of scientific investigation. See Chapter 2, n. 34.

[15] So Cajetanus, *De Nominum Analogia* 2. 10: 'analogia ista sit secundum denominationem extrinsecam tantum; ita quod primum analogatorum tantum est tale formaliter, caetera autem denominantur talia extrinsece. Sanum enim ipsum animal formaliter est, urino vero, medicina et alia huiusmodi, sana denominantur, non a sanitate eis inhaerente, sed extrinsinsece, ab alla animalis sanitate, significative vero causaliter, vel aliuo modo.'

We should not want R to permit relations grounded in purely intrinsic denomination. Aristotle's examples surely proceed by specifying relations grounded in extrinsic denominations. Thus, healthy hair is not itself healthy; it is not even alive. Still, it is called healthy because it indicates health or has an appearance characteristically caused by the condition of a generally healthy subject. Relying solely on Aristotle's examples, then, we should not permit formal causal relations to count as sanctioned by FCCP. Perhaps, then, what is wanted is not FCCP at all, but a three-causal principle.

This conclusion is mistaken, because it rests on an unduly narrow conception of formal causation. It is true that none of Aristotle's examples of core-dependent homonymy is obviously a case of formal causation. It is also true that most instances of formal causation are best understood as formal relations between intrinsically denominated subjects. Even so, an argument to show that FCCP cannot embrace formal causal relations rests upon the premise that formal causation is sufficient for univocity and so incompatible with homonymy. And this premise is false.

Although in some cases x's being F is a formal cause of y's being F only if x and y are univocally F, Aristotle also recognizes instances of formal causation where this does not hold. For example, in his philosophy of perception, Aristotle holds that every instance of perceiving consists in a sensory faculty's receiving a form of the object of perception (*aisthêton*) without its matter (*An.* 424[a]18–24, 424[a]32–[b]3, 425[b]23, 431[b]28–432[a]2). In these cases, Aristotle does not mean that the sense organs take on the perceptual qualities of the objects in question, that the organs themselves become red or square or malodorous, but that they become isomorphic with the objects of perception representatively.[16] We may say that the perceptual faculties *encode* perceptual qualities without *exemplifying* them.[17] That is, the perceptual faculties, through a system of representation, are in a state corresponding to—and fully representative of—a given set of perceptual qualities without manifesting those same qualities. When I perceive the man in the yellow hat, although I undergo a kind of process of form absorption,[18] my sensory organs do not themselves become yellow. And they do not themselves wear a hat.

---

[16] For an account of literal perception, see Sorabji 1971 and 1974. I develop the account of perceptual isomorphism and intentionality in Shields 1995. For one approach to a distinction between ways of realizing forms in perception and thought, see Brentano 1867/1975.

[17] On encoding and exemplifying, see Shields 1995 and Zalta 1988, 15–19.

[18] Aristotle believes that perception involves a kind of process incorporating alteration, but he is appropriately guarded about the status and nature of the process (*An.* 417[a]30–[b]6, 417[a]14–17, 431[a]4–7). His caution results in part from a desire not to permit the formal caus-

Since Aristotle recognizes encoding as a kind of formal realization, he does not hold that every instance of formal causation is also an instance of synonymous predication. For if someone paints a board red while another watches, we can say both:

(8) The board takes on the form of redness.

and

(9) The perceiver takes on the form of redness.

But in so speaking we ascribe importantly different processes of form absorption to the subjects. Indeed, perhaps for shorthand, we may wish to rewrite (8) and (9) as Aristotle sometimes does:

(10) The board becomes red.

and

(11) The perceiver becomes red.

where these are potentially misleading ways of expressing the observation that the perceiver and the object are 'one in form' (*An.* 402ᵇ3, 411ᵇ21, 429ᵇ28, 433ᵇ10). In these cases, it is correct to say that both the board and perceiver take on the form of redness in an intrinsic way; but it should not be inferred that they are red in the same way.

Because of the possibility of distinct types of formal realization, it is possible for Aristotle to reject any direct inference from formal sameness to synonymy. It is therefore possible for him to hold that *a* and *b* are *F*, that *a*'s being *F* is a formal cause of *b*'s being *F*, and for *a* and *b* nevertheless to be non-synonymously *F*. It is, consequently, possible for *a*'s being *F* to be a formal cause of *b*'s being *F*, even while *a* and *b* are homonymously *F*.

That said, it is difficult to produce uncontroversial examples of core and derived homonyms related by formal causation. Aristotle nowhere explicitly holds, for example, that perceivers and objects of perception are homonymously *F*. Nor can one point to cases of goodness or being as illustrations, since their non-univocity is assailable.[19] Perhaps Aristotle's difficult suggestion that some things count as medical because of their being 'well-constituted relative to it' (*tô(i) euphues einai pros autên*; *Met.*

---

ation involved in perception, which involves an active potentiality and the encoding of a property, to be confused with the formal causation involved e.g. in the painting of a board, which involves a passive potentiality and an exemplifying of the property.

[19] See Chapters 8 and 9.

1003^b2–3) is meant to provide such an illustration. If he means that non-experts with a knack for healing are called medical even though they lack the appropriate craft training, or that someone has a strong native capacity for healing,[20] then Aristotle's point will be that someone realizes the form in question incompletely or inchoately, or at any rate differently from the way a trained physician manifests it. If so, then a folk healer will count as medical because of standing in an appropriate formal causal relation to medicine.

This prospect raises an important question for our understanding of homonymy, one which asserts itself in a forceful way in discussions of the homonymy of goodness. For so far we have seen that it may be possible for two entities to count as *F*s in different ways, the first by exemplifying *F*-ness and the other by encoding it. Now the question arises as to whether two things can count as *F*s by realizing *F*-ness, not merely because one encodes what the other exemplifies, but because they exemplify *F*-ness differently. Perhaps a physician and a folk healer both stand in a formal causal relationship to medicine; but because they exemplify the medical craft in very different ways, Aristotle may want to insist that 'medical' is said of them non-univocally. Similarly, perhaps a Buddhist monk, the chief mechanic in a garage, and experience are all teachers; but to the degree that we regard them as engaging in radically different activities, or as being disparate sorts of entities altogether, we may be disinclined to treat them as exemplifying the same craft in the same way.

However that may be, it seems clear minimally that Aristotle expects some non-core homonyms to stand in formal causal relations to core homonyms. Further, he is justified in assuming that formal causation is not by itself sufficient for univocity.

We can therefore add formal causation to material, formal, and final causation as an instance of the kind of relation Aristotle expects core-dependent homonyms to realize towards one another. With this, I conclude the first two phases of my defence of Cajetan's proposal, as restructured into FCCP. FCCP grows naturally out of Aristotle's own illustrations of core-dependent homonymy; and the initial problems with FCCP in the end provide additional warrant for our endorsing Cajetan's proposal.

We are, accordingly, in a position to augment our account of core-dependent homonymy. Substituting the relevant portion of FCCP for CDH_2 (iii) yields a third, improved principle:

CDH_3: *a* and *b* are homonymously *F* in a core-dependent way iff:

[20] See n. 10 above.

(i) they have their name in common, (ii) their definitions do not completely overlap, and (iii) necessarily, if *a* is a core instance of *F*-ness, then *b*'s being *F* stands in one of the four causal relations to *a*'s being *F*.

This account is intended to capture the spirit of Cajetan's proposal by restricting the relation *R* to the four-causal scheme.

I complete my initial defence of CDH₃, and so of FCCP, by showing how it is both suitably determinate and appropriately open-textured. This is most easily appreciated by seeing that it screens unwanted junk relations. Two sorts of examples were introduced. The first attempted to portray clearly discrete homonyms as core-dependent by relying on ad hoc relations; the second attempted to extend core-dependent homonyms in questionable directions. We now see that river banks and savings banks are not core-dependent, even if all savings banks stand in the non-contingent relation 'being within five hundred miles of' a river. For though a genuine relation, 'being within five hundred miles of' is not an instance of any one of the four causes. Since *R* as specified requires this, the relation does not meet FCCP and so does not satisfy CDH₃ (iii). Hence, river banks and savings banks do not qualify as core-dependent homonyms.

Much the same can be said for 'healthy' in 'healthy salary' and 'healthy appetite'. These cases attempted to extend bona fide core-dependent homonyms beyond comfort. Now the cases can be decided by appealing to CDH₃. In so far as neither stands in one of the four causal relations to the core notion of health, neither qualifies as an instance of a derived core-dependent homonym. At the same time, these sorts of cases equally illustrate a virtue of CDH₃, namely, its open-endedness. If it turns out in a particular case that a healthy appetite is, for example, efficiently causally related to a subject's being healthy, it can be admitted as an core-dependent homonym. This does nothing to undermine the force of CDH₃. On the contrary, it permits determination on a case-by-case basis, and it does not preclude the development or extension of a homonymous concept through time. We expect our account of association to welcome new cases. CDH₃ does so by leaving open the possibility that derived instances may come to fall under FCCP. At the same time, it provides a definite and determinate decision procedure for assessing whether some concept qualifies as a core-dependent homonym.

CDH₃ provides an abstract characterization of Aristotle's illustrations of core-dependent homonymy. Importantly, it does so in a way which captures the inherent flexibility of Aristotle's notion without sacrificing the determinateness required for cleaving off unwanted relations. Even so, CDH₃ may not seem problem-free.

To begin, CDH₃ may seem not to have solved the problem of profligacy. That is, it might seem to treat clear cases of univocity as core-dependent homonyms. These would include cases recognized by Aristotle to be univocal. He holds, for example, that a father is the formal and efficient cause of a child's being a human being (*Phys.* 198ª22–32). But then a father and his child stand in the sorts of relations FCCP accepts as the appropriate sort for core-dependent homonymy. Since a father and a child are synonymously human beings, CDH₃ is incorrect.

This objection is misguided, but it does help highlight an important feature of CDH₃. It is misguided because it implicitly treats FCCP in isolation from the first two clauses of CDH₃ by proceeding as if CDH₃ (iii) were alone sufficient for core-dependent homonymy. It is not. CDH₃ (ii) requires that the cases in consideration already be non-univocal. Hence, it is not the case that any given instance of efficient causation will provide grounds for core-dependent homonymy. CDH₃ (iii) already presumes a logically prior non-univocity.

A related case is more complicated. Suppose Praxiteles sculpts a statue of a man. According to the general tests developed for non-univocity,[21] 'man' is predicated non-univocally of Praxiteles and the sculpture. Hence, this case sidesteps the rejoinder to the first objection: the logically prior requirement of non-univocity is met. Should 'man' be treated as a core-dependent homonym as applied to Praxiteles and his work?

Perhaps this consequence will be viewed favourably by some, who wish to treat these sorts of cases as instances of core-dependent homonymy.[22] This seems unwelcome, however, in so far as Aristotle's standard locutions derogate such homonyms as *mere* homonyms,[23] where the suggestion lies

---

[21] See 1. 4 and 2. 2.

[22] Whiting (1992, 77 n. 9 and 81 n. 20) raises some reasonable questions about this possibility. As I have indicated earlier (Chapter 1, n. 11), the problem results from the fact that cases of resemblance or other forms of similarity provide special circumstances. Homonymy is, however, always predicate relative. Thus, an account of the predicate 'being a sculpture of a man' will make reference to 'being a man'; but 'man' as applied to the sculpture will either be understood as elliptical for 'being a sculpture of a man' or will be an instance of discrete homonymy. The same response is available for a slight variation on our putative counter-example. If Praxiteles sculpts a sculpture of a sculptor, then we have either a case of discrete homonymy or a distinct predicate, namely 'being a sculpture of a sculptor'. Further, we can note in this connection that 'being similar to' is not sufficient for representation. Genuine representation, especially of the form we have considered, imposes requirements beyond those in place for mere similarity. A mountain range called 'the sleeping giant' because of its being similar in outline to that of a sleeping giant does not thereby represent a sleeping giant.

[23] In such contexts, Aristotle often says that an *F* is not an *F* at all, except homonymously (*plên* or *all ê homônumôs*). Some relevant passages include: *An.* 412ᵇ17–22; *Meteor.* 389ᵇ20–390ª16; *Pol.* 1253ª20–5; *PA* 640ᵇ30–641ª6; *GA* 734ᵇ25–7.

near that they are not related by the tight, informative connections he introduces for core-dependent homonyms. His language in the contexts suggests that Aristotle does not want to treat these sorts of cases as core-dependent homonyms. Fortunately, nothing about CDH_3 constrains him to do so.

Although a sculptor and his work seem to satisfy all three clauses of CDH_3, a closer appreciation of the last clause shows that this is not so. CDH_3 does not hold merely that whenever we have two non-univocal *F*s and one stands in one of the four causal relations to the others, we have core-dependent homonymy. Rather, CDH_3 (iii) specifies more narrowly that *a*'s being *F* must stand in one of the four causal relations to *b*'s being *F*. This introduces a fine-grainedness the putative counter-example neglects. For the sculptor's being a man does not efficiently cause the statue's being a man, except perhaps coincidentally. Rather, the man's being a sculptor causes the matter to be a statue of a man.

Aristotle expresses this point by distinguishing between intrinsic (*kath' hauto*) and coincidental (*kata sumbebêkota*) causes. Thus, in his discussion of chance and spontaneity in the *Physics*, Aristotle claims:

> We say that these sorts of things <viz. things which occur for the sake of something>, whenever they occur coincidentally, are by chance. For just as among what exists some things are intrinsically and others are coincidentally, so too can a cause be this way. For example, the one skilful in building is the cause of a house intrinsically, whereas the white one and the musical one are causes coincidentally. The intrinsic cause is, then, definite, whereas coincidental causes are indeterminate; for in any one case, these would be boundless. (*Phys.* 196^{b}23–9; cf. *Phys.* 197^{a}12–15; *Met.* 1032^{b}21–30, 1034^{a}9–^{b}7; *EN* 1096^{a}21)

Especially significant is Aristotle's inference that coincidental causes are indeterminate and limitless in number. His point is that for any two co-referential expressions, if the one designates an intrinsic cause, the other will designate a cause too, though it may be a coincidence that the entity so described is a cause of the result mentioned. It may be true that the pianist caused the cake to come to be; but he did not cause it qua pianist. Since the baker may equally be designated by any number of other co-referential descriptions, the cause of the cake's coming to be, considered coincidentally, is indefinite.

By contrast, what makes a cause an intrinsic cause is the availability of an appropriate connection between the properties of the events designated. Perhaps this will be an analytic connection; or perhaps it will be more weakly nomic. In either case, we see that Aristotle is aware of the fact that not every cause is an intrinsic cause. CDH_3 (iii) is intended to reflect this awareness by restricting the sorts of causal connections appropriate for core-dependent homonymy to those which are intrinsic. And while it is true that a man, a sculptor, is a coincidental cause of a man, the sculpture, this is not a case in which a man is an intrinsic cause of a man. Consequently, we do not have an instance of a case which meets CDH_3 without producing core-dependent homonymy. Hence, CDH_3 thus far remains unproblematic.

There remains, however, one serious problem with CDH_3. In surveying the kinds of illustrations of core-dependent homonymy Aristotle offers, we discovered that FCCP obtains, but only because the causal connections can head in either direction. (A regimen is an efficient cause of health, while Socrates' health is an efficient cause of his healthy complexion.) Consequently, CDH_3 does not yet capture one feature of Aristotle's commitment to core-dependent homonymy, namely, the asymmetry in dependence that derived instances bear towards core instances. What makes core instances core is that they must be cited in the accounts of all derived homonyms. But the converse does not hold true. Core instances need make no appeal to derived instances. Because it does not capture this definitional asymmetry, CDH_3 is incomplete as it stands.

4.5 DEFINITIONAL PRIORITY

CDH_3 is incomplete but not therefore incorrect. CDH_3 satisfactorily distinguishes core-dependent from discrete homonyms; it also provides a way of delimiting the sorts of relations derived homonyms may bear to core homonyms without requiring a general analysis of those relations as first-order relations. Since Aristotle offers an open-ended, motley group of relations with no obvious communality, this result too is welcome. The chore, then, is to find some additional clause, compatible with CDH_3 (i)–(iii), which provides the wanted principle of ordering.

Aristotle calls the core cases of homonymy *archai* (*Met.* 1003ᵇ6), variously rendered as 'source', 'principle', or 'first principle'. It is easy to see why he should regard core homonyms as sources of derived homonyms: derived homonyms are core-dependent homonyms precisely because they make ineliminable reference to core cases, whose accounts need contain no reference to them. It is less easy to capture the asymmetry he presupposes. Again, since the relations of derivation form no natural kind, there is no common analysis of them to yield some one form of asymmetry. The relations to which Aristotle appeals are not, for instance, uniformly causal in one direction; hence, there is no causal asymmetry between core and non-core cases. On the contrary, whereas healthy regimens cause healthy bodies, healthy complexions are indicative of health just because healthy bodies occasion healthy complexions.

None the less, Aristotle presumes some form of priority for core cases, evidently some form of definitional priority. Definitional priority need not itself be a causal or quasi-causal notion, since requirements for definitional priority are determined not by causal relations but by features of Aristotelian essentialism. Hence, we see that some F is a derived core-dependent homonym only if: (i) there is some core instance of being F; (ii) its account makes essential reference to that core instance; and (iii) an account of the core instance makes no essential reference to it. Since requirements for definitional priority admit other forms of priority beyond the merely causal, definitional priority need not be a causal or quasi-causal notion.

Aristotle recognizes many forms of priority.[24] *Categories* 12 first announces that 'one thing is called prior to another in four ways' (14ᵃ26), and he later adds a fifth, since 'there would seem to be another form of priority beyond those mentioned' (14ᵇ10–11; cf. *Met.* v. 11). The first four are relatively straightforward (*Cat.* 14ᵃ26–ᵇ8): (i) priority in time; (ii) priority as regards implication of existence;[25] (iii) priority in order (e.g. in speeches or presentations); and (iv) priority in value (as in 'They give priority to members of their own family'). The fifth form of priority is somewhat more difficult to characterize. Between two things which reciprocate as regards implication of existence, one may nevertheless 'in some way be the cause of the existence of the other' (*to aition hopôsoun thaterô(i) tou einai*; 14ᵇ12) and so may 'reasonably be called prior by nature' (*proteron eikotôs phusei legoit' an*; 14ᵇ12–13). Aristotle's idea here is somewhat unclear and may be unfortunately expressed.[26] The sort of example he offers helps somewhat. Socrates' being white and the true proposition (*logos*) that Socrates is white may reciprocate as regards implication of existence, even though Socrates' being white is responsible for the proposition's being true, whereas its being true, by contrast, is not responsible for Socrates' being white (cf. *Met.* 1051ᵇ6–8).[27]

Both the second and fifth forms of priority have the asymmetry required for a refinement of CDH₃. The first form of priority has the wanted asymmetry but is clearly not relevant; the third form is possibly asymmetrical but also not relevant; the fourth kind is not asymmetrical in the required way. The second form has the wanted asymmetry in so far as

---

[24] On forms of priority, see Cleary 1988.

[25] At *Metaphysics* 1019ᵃ4, Aristotle credits Plato with having used this sense of priority.

[26] See Ackrill 1963, 111–12.

[27] This explanation takes some liberties with Aristotle's actual language; he says that Socrates' being white is responsible not for the proposition's being true, but for its very existence (*to aition hopôsoun thaterô(i) tou einai*; *Cat.* 14ᵇ14).

it allows that *F*s are prior to *G*s just in case the existence of *G*s entail the existence of *F*s but not vice versa. Non-core core-dependent homonyms require the existence of core cases; it is less clear whether core cases entail the existence of non-core cases. The question of whether the second or fifth form of priority, if either, provides the form required by CDH_3 turns in part on whether core homonyms require the existence or merely the possible existence of non-core cases.

Perhaps an example will help make this point clear. If 'healthy' in 'healthy complexions' and 'healthy people' is homonymous, there will be some asymmetry between them. Healthy complexions are healthy only because they indicate health in the people who sport them. (The features we associate with healthy complexions could as a nomological possibility have been indicative of illness and would not then have been healthy complexions.) Healthy people remain healthy even when their health is not visibly manifested in their complexions. More to the point, the account of 'healthy' in 'healthy complexion' will refer to health, whereas the account of health as such will not refer to complexions.

Still, if people are healthy, there will likely be some state or other indicative of health, some practices productive of health, and so forth. There need not be some one state or some one practice; rather, there must be some one category, being indicative of health or being productive of health, such that one or more states can play that role. This suggests that there may be some necessary connections between the existence of core homonyms and the existence of non-core homonyms: core homonyms are responsible for the existence of non-core members of the functionally determined categories whose principal instances they are.

If this is correct, then Aristotle's fifth type of priority may help capture the form of asymmetry required. Core and non-core homonyms may reciprocate as regards implication of existence, even though core homonyms are responsible for the existence of non-core homonyms in a way that non-core homonyms are not responsible for the existence of the core cases. Within the framework of the fifth type of priority mentioned in the *Categories*, a core homonym will be in a way the cause of the existence of a derived homonym, or perhaps more broadly will be *responsible* for its existence (*to aition hopôsoun thaterô(i) tou einai*). This points towards an emendation to CDH_3 which yields:

> CDH_4: *a* and *b* are homonymously *F* in a core-dependent way iff: (i) they have their name in common, (ii) their definitions do not completely overlap, (iii) necessarily, if *a* is a core instance of *F*-ness, then *b*'s being *F* stands in one of the four causal relations to *a*'s being *F*,

and (iv) a's being F is asymmetrically responsible for the existence of b's being F.[28]

On this approach, the key to Aristotle's view of core-dependence lies in the sense in which x can be responsible for the existence of y, even when x and y reciprocate as regards implication of existence. Evidently, in suggesting that one thing can be responsible for the being of another thing *as an F*, Aristotle means to suggest that its being F is itself derived from the source's character. A state of affairs is responsible for a proposition's being true precisely because the state of affairs is, so to speak, the truth-maker. Although a true proposition p and a state of affairs may exist only if the other does, the proposition's being true results from its agreement with the structure of the state of affairs, and not conversely. Here, then, is a primitive form of priority, not reducible to any other form, which Aristotle may intend in calling the core instances *archai*.

If core homonyms are prior in this sense, then there is some constraint on the kind of *archê*, or source, core homonyms can be. This is fortunate, since the notion of an *archê* is itself an open-textured one. In *Metaphysics* v. 1, Aristotle retails at least six different senses of *archê*, only some of which are relevant to his conception of priority in account. For not all kinds of *archai* display the kind of priority we expect homonyms to display.[29] One sort he mentions is, however, directly relevant. One thing can be the *archê* of another in the sense of being 'that from which something can be first known' (*Met.* 1013ᵃ14–15), where the illustration is that of a premise in a demonstration.[30] Premises are the sources of their conclusion in the sense that they are responsible for the existence of a proposition's being a conclusion, something entailed by another set of propositions. In this sense, a proposition's status as a conclusion derives from its premises because of primitive entailment relations, which are not reducible to

---

[28] This account presumes that one of the two associated homonyms is in fact the core. This, as a formal matter, need not be the case. Thus, one could rewrite CDH₄ more cumbersomely, as follows, to reflect this fact: CDH₄*: x and y are homonymously F in a core-dependent way iff: (i) they have their name in common; (ii) their definitions do not completely overlap; (iii) necessarily, if x is a core instance of F-ness, then y's being F stands in one of the four causal relations to x's being F; and (iv) (a) x is asymmetrically responsible for the existence of y, or (b) y is asymmetrically responsible for the existence of x, or (c) there is some z that is asymmetrically responsible for the existence of both.

[29] Among the inappropriate forms are: (i) a beginning, in sense of the first part of a road traversed, (ii) the first part completed, as in the keel of a ship, and (iii) that from which a natural progression proceeds, as abusive language leads to fighting (*Met.* 1012ᵇ32–1013ᵃ10).

[30] I agree with Ross (1924, i. 291) that the hypotheses of *tôn apodeixeôn hai hupotheseis* at 1013ᵃ16 are to be understood 'in the general sense of "premises"', as at 1013ᵇ24.

any form of derivation. So too is the asymmetry in core-dependent homonymy primitive: the cores are sources because they are semantically and metaphysically super-ordinate to their derivations.[31] A complexion is healthy because its being healthy is explicable in no other way than in its standing in an appropriate relation to a healthy entity.

If this is correct, the form of asymmetry required for core-dependent homonyms is primitive. It cannot be reduced to some other form of priority. Even so, it can be explicated by placing it in a framework of similar cases, including the form of priority we find in premises and conclusions. While the soundness of an argument guarantees the truth of both its premises and its conclusion, it is the premises which entail the conclusion and not the other way around.

4.6 CONCLUSIONS

Aristotle's examples of core-dependent homonymy are intended to illustrate a general form of unity in multiplicity which some, but not all, non-univocal terms enjoy. By extrapolating from these examples, it is possible to develop a general account of core-dependent homonymy along the lines of CDH_4. According to CDH_4, core-dependent homonyms (i) are non-univocal, because their accounts are discrete; (ii) have core and non-core instances whose relations are constrained by Aristotle's four-causal scheme; and (iii) have core instances whose accounts are necessarily cited in the accounts of all derived instances, because they are *archai* of the derived instances in the sense of being asymmetrically prior to them. This general framework captures both the determinateness and the open-endedness of Aristotle's approach. Core-dependent homonyms are determinate, because every derived instance must stand in one of the four causal relations to the core. Still, it is an open question as to the variety of first-order relations core and non-core relations may exhibit. So long as they are related by one of the four causes, they will be candidates for core-dependent homonymy.

It should be clear that CDH_4 will not by itself determine whether any given philosophical concept of significance is a core-dependent homonym. Again, the general framework is a framework and nothing more. It is not a decision procedure; and it does not by itself arbitrate disputed

---

[31] The notion of super-ordination here is roughly that of Katz (1990, 1994), who exploits the notion of super-ordination and subordination in a free-standing intensionalist semantic structure.

cases. Only analysis yields this result. Still, we are now in a position to move from the relatively uncontroversial illustrations Aristotle uses to motivate his doctrine of homonymy to the controversial cases these examples are intended to illustrate. I turn now, in Part II, to an examination of homonymy at work.

II

Homonymy at Work

5

The Body

5. 1 A DISCRETE, SEDUCTIVE HOMONYM

Aristotle's distinction between seductive and non-seductive homonymy cuts across his distinction between discrete and associated homonymy.[1] Hence, among the varieties of homonymy, Aristotle recognizes a category which is both discrete and seductive. This may be in some ways surprising, since we do not expect discrete homonyms to be seductive. In general, if the occurrences of *F* in '*a* is *F*' and '*b* is *F*' are homonymous but not associated, the distinctions between them will be obvious, and available to any competent speaker. Aristotle's introduction of discrete homonyms which are not immediately obvious therefore requires special pleading.

In this chapter I justify Aristotle's introduction of discrete seductive homonymy by considering its principal application: the body. I argue that the body provides a defensible application of homonymy, one which helps explain why Aristotle would be motivated to recognize such a category in the first place. My overarching aim, however, is not narrowly taxonomical. Rather, I wish to show how Aristotle's appeals to homonymy help stave off what would otherwise be cogent criticisms of his hylomorphism in philosophy of mind. If my defence of Aristotle against these criticisms is successful, and if, as I argue, 'body' is a discrete homonym, then my remarks also justify Aristotle's introduction of seductive discrete homonymy. The special pleading will be direct: philosophical contexts sometimes demand defensible distinctions to which everyday contexts can be safely indifferent. Indeed, because they will arise only in philosophical disputation, such distinctions may seem positively peculiar from a prosaic point of view. This is why even discrete homonymy may be seductive.

5. 2 HOMONYMY AND HYLOMORPHISM

Aristotle contends that the body is homonymous (*An.* 412[b]20–5, 412[b]27–413[a]2), and he supposes this claim to be of service in his

---

[1] See 1. 5 and 1. 6.

hylomorphic analysis of soul and body. He suggests, in fact, that a dead body is not a body *except* homonymously (*plên homônumôs*) and evidently takes this to shore up his contentions that human beings have characteristic functions (*erga*) of a determinative sort and that human bodies have these functions in virtue of being ensouled. His view would seem to be that a functionally specified human body must be able to engage in complicated sorts of activities, including nutrition and perception, and that dead bodies fail to qualify for the simple reason that they lack these abilities. As far as this observation goes, we might initially suppose Aristotle justified in suggesting that Perikles' corpse has more in common with a statue of Perikles than it does with the body of the living man: the statue and the corpse may both equally bear some superficial resemblance to Perikles, but neither could be expected to share in even his most basic abilities. This is no doubt why, in the old Oxford translation of *plên homônumôs*, Aristotle would maintain that a dead body is not a body 'except in name'.[2]

From this perspective the homonymy of the body appears to dovetail rather nicely with other features of Aristotle's hylomorphic analysis of soul and body. But this appearance, like the old Oxford translation, merely obscures deeper difficulties: there are problems with Aristotle's conception of the homonymy of the body. Some of these problems might be thought counter-intuitive but tolerable;[3] others appear problematic only for commentators advancing functionalist interpretations of Aristotle's philosophy of mind;[4] and as Ackrill has argued in an influential article,[5] still others are quite general and evidently threaten the intelligibility of Aristotle's entire hylomorphic project in the philosophy of mind.

To begin, it would seem that if the body is homonymous, and if dead bodies are not really human bodies at all, human bodies, no less than human souls, go out of existence at the moment of death. We have imme-

---

[2] See the Oxford translation of J. A. Smith, *An.* 412ᵇ21. This translation is somewhat surprisingly retained in the Revised Oxford Translation (Barnes 1984).

[3] Some objections raised by Williams (1986) fall into this category.

[4] In Shields 1990 I argue that Aristotle is a functionalist and develop the precise commitments of his theory. For further discussions of functionalism in Aristotle see Nussbaum and Putnam 1992; Hartman 1977, ch. 4 § V; Robinson 1978; Modrak 1987, chs. 1 and 2; Cohen 1987; and Nussbaum 1978.

[5] Ackrill 1972–3/1978; references are to the latter pagination. See also Burnyeat 1992. Burnyeat specifically argues along the lines developed by Ackrill in his attack on the intelligibility of Aristotle's philosophy of mind. He also has other objections directed against functionalist interpretations of Aristotle and indeed against any account which regards his views as anything more than a primitive, explanatorily inefficacious form of hylozoism. These objections stem from a detailed analysis of Aristotle's account of perception (*aisthêsis*) and do not depend essentially upon an appeal to the homonymy principle; consequently, I will not undertake to discuss them in the present context. They are in any case adequately refuted by Cohen (1987).

diately, then, the counter-intuitive result that there is evidently no continuing subject called the 'human body' which persists through the death
of the organism. But if the human body does not survive the death of the
organism, we speak falsely when we say, for example, 'I went to Red
Square and saw Lenin's body'. Lenin's body no longer exists, and what we
say is neither more nor less apt than if we had uttered the same phrase having seen a statue or even a picture of Lenin. In this instance, it is not at all
clear what relationship the lump of stuff in Red Square bears to Lenin or
even to Lenin's body, if any at all. Given the homonymy principle, there is
prima facie no more perspicuous connection between the matter of
Lenin's corpse and Lenin's body than there is between a statue of Lenin
and Lenin's body.

Moreover, given that he evidently regards the bodies of all ensouled
creatures as homonymous, including those of animals and plants,
Aristotle's point is really quite general. Just as a dead human body is no
longer a human body and cannot even be regarded as a dead human body,
so the body of a dead squirrel is no longer a squirrel body, and the body—
if so it may be called—of a corn stalk is no longer a corn stalk body once
the plant is uprooted. Put this way, Aristotle's doctrine begins to look
somewhat perverse: he evidently maintains that what one holds in one's
hands after uprooting a corn stalk is in some real sense no longer the same
body one held before uprooting it. And this surely sounds counter-
intuitive.

Indeed, one might think this result especially counter-intuitive, or
worse, when considered against the backdrop of Aristotle's hylomorphism. If the soul–body relationship is, as Aristotle asserts, a special case
of the form–matter relationship (*An.* 412ᵃ19–21, cf. 412ᵇ6–9), then we
would expect the body, as matter, to have modal properties distinct from
the form, in the way in which the bronze in the statue of Hermes has the
modal property of being potentially the bronze in the statue of Theseus,
something the form of the statue of Hermes clearly lacks. But if human
bodies go out of existence at the instant of death, they are ill suited to be
the matter of other entities. If so, the claims that the soul is the form of
the body and the body the matter of the soul begin to strain the notions
of form and matter almost beyond recognition by failing to sustain the
very analysis of change and generation for which they were initially introduced in *Physics* i.[6]

---

[6] See Ackrill 1972–3/1978, 69–70: 'The problem with Aristotle's application of the
matter–form distinction to living things is that the body that is here the matter is itself
"already" necessarily living . . . the material in this case is *not* capable of existing *except* as
the material of an animal, as matter *so in-formed*' (italics as found). Ackrill succinctly notes
that organs and bodies 'are necessarily actually alive' (70). If this is right, then it will be

This general problem may seem especially pointed for functionalist interpretations of Aristotle's psychology. Two distinct but related features of Aristotle's claim that the body is homonymous appear to be incompatible with functionalist accounts of Aristotelian psychic states: (1) functional states are contingently related to material states, while the homonymy principle as applied to bodies suggests a stronger connection; and (2) if functionalism is correct, mental states are multiply realizable, but it is unclear how Aristotle could regard them as such given his commitment to the homonymy of the body. Only bodies of a certain sort, viz. bodies that are already living, are appropriately regarded as potential bases for realizing mental states; if non-living bodies are not even potentially human bodies, then it is unclear how Aristotle could endorse any sort of multiple realizability for mental states.

In spite of these prima-facie difficulties, Aristotle's contention that the body is homonymous can be explicated within the framework of hylomorphism, and to a certain extent it can be used to defend that framework from objections some commentators have found compelling. In some cases, problems commentators have located in Aristotle's doctrine result from underdeveloped appreciations of the conceptual underpinnings of Aristotle's commitment to homonymy; in others, the results offered as counter-intuitive strengthen rather than militate against the homonymy of the body. Moreover, once understood, it becomes clear that far from undermining functionalist interpretations, Aristotle's doctrine of the homonymy of the body coheres perfectly with this approach.

5.3 DISCRETELY HOMONYMOUS BODIES

Aristotle's claim that the body is homonymous is initially striking, because he treats living and dead bodies as discretely homonymous. Although, as we have seen,[7] Aristotle accepts CH, or comprehensive homonymy, as his general thesis, CH embraces those homonyms recognized by DH, or discrete homonymy, as special cases. If he thinks that bodies are discretely homonymous, Aristotle will maintain that the definitions of living bodies have nothing in common at all with those of corpses, and this may seem somewhat startling.

Now, however, it needs to be appreciated that some of the very passages

---

difficult to understand how some matter can serve as the continuing substrate in generation and destruction. For a clear discussion of some of these issues, see Lewis 1994, 249–54 and 268–72.

[7] See I. 1–I. 3.

which support attributing CH to Aristotle also show that he regards living and dead bodies as *mere* homonyms (*An.* 412ᵇ20–5), that is, as homonyms which are non-associated and thus not core-dependent. Aristotle regularly places the body and bodily parts into the category of non-associated homonymy by aligning them with other sorts of non-associated homonyms, including paintings and sculptures (*An.* 412ᵇ17–22, *Meteor.* 389ᵇ20–390ᵃ16, *Pol.* 1253ᵃ20–5, *PA* 640ᵇ30–641ᵃ6, *GA* 734ᵇ25–7). In these contexts, he often uses this locution: *x* is not an *F*, or *x* is no longer an *F*, 'except homonymously' (*plên homônumôs*; e.g. *An.* 412ᵇ21; he sometimes makes the same point by saying 'or [is] rather [an *F*] homonymously', *all' ê homônumôs*, e.g. *An.* 412ᵇ14–15). In so speaking, he means that the *F*s in question have nothing definitionally in common with genuine *F*s and are called *F*s only by custom or courtesy. Thus, he recognizes the bodies as in the sub-class of homonyms, discrete homonyms.

What is Aristotle's motivation for holding this strong view? When it is pointed out that the homonymy principle as applied to bodies produces untoward results, why does he not simply dilute it in order to avoid this difficulty?[8] Conversely, if he sees that his commitment to the homonymy of the body has uncomfortable results, why does he not simply change his attitude towards dead bodies and suppose that they are real bodies, except that they are no longer animated?

The problem with both solutions is the same. Neither appreciates that Aristotle cannot alter the homonymy principle or restrict its application at will. For as we have seen, in establishing non-univocity, Aristotle relies on his functional analysis of kind identification, according to which an individual *x* will belong to a kind or class *F* iff *x* can perform the function of that kind or class.[9] Thus, according to FD, it is both necessary and sufficient for *x*'s being a member of kind *F* that *x* have the functional capacity determinative of *F*s. Given that ensouled and non-ensouled bodies will trivially have different sets of capacities, FD in turn entails that they be treated as non-univocal. It is therefore not open to Aristotle simply to reject or restrict his commitment to homonymy in such a way that bodies not be treated homonymously. To do so, he would first need to reject a pervasive and broad metaphysical commitment.

When he says that dead bodies are bodies only homonymously, he indicates that dead bodies, living bodies, and representations of both form a class worthy of our attention, even though—or rather because—they have

---

[8] Thus, Ackrill (1972–3/1978, 71) offers Aristotle a weakened homonymy principle, but rightly adds that 'this suggestion does not get to the root of the problem'. I sketch the reasons why it does not in this section.

[9] See 1. 5 on the functional determination thesis.

nothing definitionally in common. In this sense, it might seem that discrete homonyms differ in kind rather than in degree from genuine homonyms. But this is misleading. Discrete homonyms have no definitional overlap, even though this may not be obvious to the untutored eye. Consequently, the philosophical danger is essentially the same in both cases: without appropriate caution, we may be inclined to classify together with genuine *F*s entities which are at best similar to *F*s in terms of their superficial characteristics. In the case of the body, living and dead bodies, as well as portraits and statues of bodies, are homonymous, because while living bodies have one sort of essence, dead bodies and statues have another or none at all.

5. 4 ORGANIC AND NON-ORGANIC BODIES

It is, however, extremely important to recognize that when he speaks of bodies as homonyms, Aristotle evidently means to differentiate not only corpses from living bodies but also one type of living body from another. Although, as I will argue, contemporary commentators who find difficulty with the application of the homonymy principle to the body have overlooked this distinction, Aristotle is careful to distinguish what he calls the 'organic' (*organikon*) body from one which is not. In his canonical definition of *psuchê* at *De Anima* 412ª19–20 he claims that it is the 'form of a body having life in potentiality' (*eidos sômatos phusikou dunamei zôên echontos*), and after clarifying that the soul is form as first actuality, he adds at 412ª28–ᵇ1, 'this sort of body would be one which is organic' (*toiouton de ho an ê(i) organikon*).[10] He never explicitly names the class of bodies distinguished from those which are called 'organic'; I will refer to them as 'non-organic bodies'.

It will be necessary to mark this distinction in part in order to heed Aristotle's advice and avoid obscuring class distinctions by failing to notice cases of homonymy. Since it seems clear that something, if not the organic body, exists both before and after the instant of death, it must be the case that before death one has a non-organic as well as an organic body.[11] Organic and non-organic bodies are to be distinguished not by

---

[10] I do not agree with Alexander of Aphrodisias, *De Anima cum Mantissa*, ll. 104.11–15, that in so speaking Aristotle identifies the organic body as the only body having life in potentiality. Rather, a body of which the soul is the first actuality has life in potentiality and is organic.

[11] This does not follow directly, but Aristotle identifies the non-organic body as one which has lost the soul (*to apobeblêkos tên psuchên*; *An.* 412ᵇ25–6), thereby implying that it

being alive or dead but by the way they are alive when they are alive. An organic body, as essentially alive, will have this property necessarily, so that in the case of human beings, Aristotle will hold:

> Organic Body (OB₁): x is the organic body of a human being iff: (i) x occupies space in itself (*kath' hauto*),[12] and (ii) necessarily x is ensouled by a human soul.

But given Aristotle's functional analysis, clause (ii) of this definition can usefully be explicated so that we have:

> Organic Body (OB₂): x is the organic body of a human being iff: (i) x occupies space in itself (*kath' hauto*), and (ii) necessarily x can perform the functions of a human being.

According to this definition, something is an organic body only if it can perform the activities functionally determinative of a human being and has this capacity necessarily. This captures Aristotle's contention that organic bodies cease to exist when they no longer have the requisite capacities, and it makes sense of Aristotle's holding, as Ackrill puts it, that bodies are 'necessarily actually alive'.[13] The class of non-organic bodies, then, do not have human capacities necessarily, but rather contingently. Hence, non-organic bodies are simply organic bodies without this modal property:

> Non-organic Body (NOB): x is the non-organic body of a human being iff: (i) x occupies space in itself (*kath' hauto*); (ii) at t_1 x can perform the functions of a human being; and (iii) at t_2 x cannot perform the functions of a human being.

On this definition, it will be true to say at some time that non-organic bodies can perform characteristically human functions; but they will not have this property essentially or necessarily. This enables them to persist beyond the destruction of the organic bodies and to be identifiable as such when not ensouled.

---

at one time had a soul. Hence, the non-organic body does not come into existence at the moment of death.

[12] Clause (i) indicates only that organic bodies occupy space. Aristotle distinguishes various types of matter including perceptible (*aisthêtê hulê*) and intelligible (*noêtê hulê*), in the *Metaphysics*, e.g. at 1025ᵇ34, 1036ᵃ9, 1037ᵃ34, and 1045ᵃ34 (on which distinction see Ross 1924, ii. 200). Aristotle takes organic bodies to be minimally perceptible, as his suggestion that mortal living things must have a capacity for motion in space (*kata topon*) indicates (*An.* 413ᵃ31–ᵇ4).

[13] Ackrill 1972–3/1978, 70.

5. 5 DISCRETE BODIES/DISCRETE PROPERTIES

It remains opaque what sort of relationship non-organic bodies will bear to organic bodies, but before attempting to specify what this might be, it will be worth showing that the distinction is serviceable and that it helps resolve the general problem of the homonymy of the body introduced by Ackrill, who summarizes his argument as follows:[14]

Aristotle's definitions of *psuchê* resist interpretation because (i) the contrast of form and matter in a composite makes ready sense only where the matter can be picked out in such a way that it could be conceived as existing without the form, but (ii) his account of body and bodily organs makes unintelligible, given the homonymy principle, the suggestion that this body or these organs might have lacked a *psuchê*. The complaint is not that Aristotle's concept of matter and form commits him to the impossible notion that what has form must lack it—that the same matter both has and has not the form; but that it commits him to something that he cannot allow to be possible in the case of living beings, namely that what has the form might have lacked it—that the same matter has and might not have had the form.

Ackrill clearly regards the homonymy principle as responsible for the difficulties he finds in explicating Aristotle's definition of *psuchê*, and he suggests that the problem has to do with the application of the principle to living entities. The problem, then, is that at t_1, whereas an enformed mass of bronze, or other material stuff, will have modal properties which its correlative form lacks at t_1, this will not be the case for living bodies. Thus, Aristotle's studied tri-part analogy

form : matter :: Hermes' shape : bronze :: soul : body

breaks down at a crucial juncture, since bodies are *necessarily* ensouled, whereas bronze is contingently Hermes-enformed.

If Aristotelian bodies, or more precisely what I have called organic bodies, are necessarily ensouled, then they will not—indeed they *cannot*—exist before or after being ensouled and will therefore be disanalogous with paradigmatic cases of Aristotelian matter, e.g. the bronze in the statue of Hermes. But why should this present a special difficulty for Aristotle? Ackrill's worry about the necessary ensoulment of Aristotelian bodies derives from two distinct quarters, and it will be useful to explicate the distinct arguments driving his concerns.

First there is a problem about generation. Given their ontological dependence on souls, organic bodies will not be able to perform the function for which matter was initially introduced, namely, to serve as the sub-

---

[14] Ackrill 1972–3/1978, 70.

strates (*hupokeimena*) underlying qualitative change and substantial generation. Organic bodies, as necessarily ensouled, cannot exist before the acquisition of the form they instantiate, and they perish when they no longer realize that form. Consequently, they cannot be the substrates persisting through substantial generation or destruction.

The second problem seems to be a deeper, conceptual one. Ackrill suggests that hylomorphism 'makes ready sense' only where matter can be picked out at t_1 in such a way that it might not have been informed as it is at t_1, viz. that matter must always have some modal property which its correlative form lacks.[15]

I will discuss each of these concerns in order to show that neither presents any problem for Aristotle and that, consequently, his doctrine of the homonymy of the body escapes the difficulties some have associated with it. Indeed, far from generating problems, the homonymy of the body helps articulate and defend the framework of hylomorphism.

In the first case, it is surely correct that organic bodies do not exist before the genesis of the organism or persist through its destruction. As necessarily alive, an organic body persists only so long as does the organism whose body it is.[16] But in holding that organic bodies are necessarily alive, Aristotle does not commit himself to the further claim that any

---

[15] A third potential problem will be that since organic bodies will not have obvious modal properties which the forms they instantiate lack, it will be difficult for Aristotle to offer a clear principle of individuation for them. In what sense, then, can he maintain that organic bodies are anything other than individual souls? Put another way, in what sense will diachronic bits of matter be anything other than particular forms? If organic bodies are taken to *be* souls, then (a) souls will need to be taken as essentially material; (b) evidently each of their capacities would have a material organ, something Aristotle clearly denies; and (c) and generally, given Leibniz's Law, everything Aristotle says of organic bodies (in non-intentional contexts) will be true of particular souls and vice versa. Some commentators will welcome these conclusions. Whiting (1986, 359–77) argues clearly and forcefully for them. I have argued against the identification of souls with bodies in Aristotle on the grounds that any such identification would involve Aristotle in an unacceptable and unnecessary violation of Leibniz's Law. See Shields 1988a. To those arguments, I can add in the present context: (1) if particular souls were identical with diachronic material continuants, then each psychic capacity would be material; (2) in *De Anima* iii. 4 Aristotle clearly denies this in the case of one psychic capacity, *nous*; therefore (3) particular souls are not identical to diachronic material continuants. As far as this argument is concerned, particular souls could still be essentially material, since they could be identical to some sort of composite entity, one component of which was essentially material (although this is also something I reject). If this were so, then the unacceptable result that organic bodies were partially immaterial entities would follow.

[16] With the possible exception of *De Anima* 407^{b}17–26, I see nothing in Aristotle's account which logically requires that an organic body be the matter for numerically one and the same soul throughout its existence. But I ignore this possibility since (a) Aristotle is best understood as holding to nomological correlations between bodily and psychic states; and (b) my present argument does not depend upon recognizing it.

potentially organic body is necessarily alive, as Ackrill seems to suggest in claiming: 'Until there is a living thing . . . there is no "body potentially alive"; and once there is, its body is necessarily actually alive.'[17] Given that the organic body is generated only when ensouled, what Ackrill says here is correct if by 'body' he means 'organic body'. But if so, it is not clear why this result should present a difficulty for Aristotle, who is willing to distinguish organic from non-organic bodies.

There would be a problem if Aristotle held an unacceptably strong principle of generation,[18] according to which the pre-existing substrate were necessarily already what he calls the *proximate matter* of the organism.[19] For example, Aristotle would be inconsistent if he held a principle of the following sort:

> Substrate as Proximate Matter (SPM): The proximate matter of x must persist (or must be able to persist) as such through the substantial generation (or destruction) of x.

If this were his view, then Aristotle would be committed to holding that the organic body both must and must not precede (and both must and must not persist beyond) its ensoulment. And this would certainly be incoherent.

When Ackrill claims that there is 'no body potentially alive', he rules out the possibility that there could be a body, not the organic body, which was potentially alive, and so implicitly presupposes such a strong principle. But without ignoring Aristotle's distinction between organic and non-organic bodies, this possibility could be ruled out only if it were true that the only potentially living body were the organic body. Such a claim in turn could be justified only by SPM. Should Aristotle endorse SPM? Is there evidence that he in fact does?

One might suppose already Aristotle constrained to hold SPM, or some formulation of it restricted to living organisms, in virtue of his commitment to the existence of organic bodies, which are after all *necessarily* alive.[20] But such a supposition would be mistaken. Four planks of wood

---

[17] Ackrill 1972–3/1978, 74.

[18] In what follows I ignore complications arising from Aristotle's analysis of the generation of organisms in terms of the paternal form informing the maternal matter.

[19] There are competing analyses of proximate matter in Aristotle. I accept Heinaman's characterization of proximate matter as the sort of matter sufficiently complex to serve as the matter of a given compound, so that e.g. the earth and water which constitute a quantity of bronze will be the non-proximate matter of a bronze statue, while the bronze itself is its proximate matter. See Heinaman 1979, 256–7.

[20] It is not entirely clear whether this is Ackrill's motivation. When he suggests that the only potentially living bodies are already necessarily alive, he seems to rely on this sort of principle.

of the same length set out vertically on a table are potentially square-shaped. If they are so arranged, they form an object which necessarily has four interior right angles equalling 360°. That squares necessarily have this property does not entail that no non-square can come to constitute a square-shaped object. This same point may be put without employing an example where the necessity is so clearly analytic. Some basic elements could come to form gold when subjected to the right environment; when they do, they will constitute something with an atomic number of seventy-nine.[21] The same may hold true of the non-organic body: a non-organic body could potentially be an organic body, in a sense to be specified, even though it is itself not necessarily alive. There would, in this case, be a substrate continuing through the substantial generation and destruction of the organism, even though it would not be the organic body, which would be potentially such as to constitute something necessarily alive.

That conceded, one will nevertheless wish to know whether there is evidence that Aristotle endorses SPM, needlessly or not. In some passages, e.g. *Metaphysics* 1032ᵃ24–ᵇ6, he does seem to endorse SPM. There he holds that a compound is generated, but not the matter or form of which it is compounded, since it is always necessary for the matter and form to pre-exist. But this passage does not require that the matter which pre-exists must be the *proximate* matter of the compound. And in any case, this passage must be considered together with Aristotle's more self-conscious accounts of generation. Both in *Physics* i. 6–8 and in the difficult discussion of *De Generatione et Corruptione* i. 3 (esp. 318ᵃ27–319ᵃ2), Aristotle affirms his commitment to the impossibility of generation *ex nihilo* even though what he calls generation *simpliciter* (*genesis haplôs*) is admitted. In these discussions he therefore commits himself to the necessity of there being a pre-existing substrate from which every substance is generated, but nowhere does he hold that the proximate matter of x must persist as such through the substantial generation of x. On the contrary, the discussion in *De Generatione et Corruptione* suggests that he rejects this principle.

Inquiring into the generation of substances[22] in this discussion,

---

[21] An example more illustrative of Aristotle's functional analysis of kind identification might be the following. A lump of iron certainly does not as such have the property of being able to cut; nevertheless, it potentially constitutes something which has this property, and has it necessarily. The necessity in these cases, as regards the constitutive stuffs, is hypothetical: this stuff necessarily has capacity G if it is to constitute something of kind F. As we shall see below, there is also a way of characterizing the matter which makes the necessity analytic.

[22] See Williams 1985, 80–96, for a discussion of some of the issues surrounding this passage.

Aristotle asks a question of 'remarkable difficulty . . . how coming to be *simpliciter*' is possible (*GC* 317ᵇ17–19). The analysis of how this is possible involves first distinguishing senses of generation *simpliciter*, and second explaining why cases of alleged generation *simpliciter* are not really instances of alteration (*alloiôsis*). In the concluding sections of the discussion, Aristotle asks of the matter serving as the substrate in particular cases of generation: 'Is there matter in one way the same, but in another way different?' He responds, 'For that which underlies (*hupokeitai*), whatever it is, is the same, but being <for it> is not the same' (*GC* 319ᵇ23–4).[23] When he uses the formula '*x* and *y* are the same, but differ in being', Aristotle usually means that *x* and *y* overlap spatiotemporally but fall under distinct sortals with non-equivalent reference classes. In the case under discussion, then, Aristotle suggests that in generation, the substrate (*hupokeimenon*) itself can alter in the course of generation, so that it comes to fall under new sortal predicates as it comes to underlie a new form. If so, then he explicitly allows cases where SPM does not obtain. Consequently, there is no reason to suppose that Aristotle needs to accept SPM or that he does in fact accept it, and indeed there is evidence to suggest that he rejects it.[24]

Especially instructive in this regard are those cases discussed by Aristotle where discretely homonymous *F*s come to constitute genuine *F*s by coming to acquire the function determinative of *F*s. An interesting and important example comes from *De Anima* i. 4, where Aristotle allows that an old man with deteriorating sight could receive an eye of an appropriate sort and thereby improve his vision (*An.* 408ᵇ20–2). In this case, Aristotle imagines an eye, indeed a discretely homonymous eye,[25] which replaces a genuine eye and thereby itself comes to constitute a genuine eye by being fitted into the nexus of a functioning organic body. As an eye, it necessarily has the property of being able to see; before being put in place, the discretely homonymous eye only potentially constituted something with that necessary property. By entering into a functioning body, the discretely homonymous eye, which is clearly the matter pre-existing in this

---

[23] Joachim (1922, 105) rightly offers the following paraphrastic translation: 'For that which underlies them, whatever its nature may be *qua* underlying them, is the same, but its actual being is not the same.'

[24] Further evidence for this same conclusion comes from *Met.* 1044ᵇ34–1045ᵃ6, where Aristotle suggests that the underlying substrate in the transition from wine to vinegar is water rather than the wine itself. In the same passage, he suggests that there is an underlying substrate in the transition from a living being to a corpse; here the only plausible candidate would seem to be the non-organic body.

[25] Aristotle himself uses the eye as an illustration of discrete homonymy at *An.* 412ᵇ20–2.

case of generation, takes on a new form and becomes an actual eye. If so, we have an instance of non-organic stuff coming to constitute organic stuff, while the latter on Aristotle's account is properly regarded as the matter of the eye. Consequently, he himself proposes a case where SPM fails where discrete homonyms are concerned.

The special case of the eye shows that Aristotle is not committed to SPM, even where living organisms are concerned. But without a commitment to SPM, the homonymy principle presents no difficulty for Aristotle's account of generation for human beings, so long as one recognizes his motivated distinction between organic and non-organic bodies. He can hold that the non-organic body (a) potentially constitutes the organic body, which is necessarily alive;[26] and (b) serves as the only substrate required in generation. Therefore, the homonymy of the body is compatible with Aristotle's account of generation and does not create any difficulty for him in this regard.

The second argument implicit in Ackrill's account of the consequences of homonymy as applied to living bodies does not depend upon Aristotle's account of generation *per se*, although if made out it could have unacceptable consequences for that account. Ackrill suggests that the homonymy principle creates a conceptual problem for Aristotle, at least in the case of the human body, since it 'makes ready sense' only where matter can be identified independently of the form whose matter it is. In paradigmatic cases, this seems easily done: the bronze in the statue of Hermes

---

[26] Some commentators have thought this possibility blocked by *An.* 412ᵇ25–6, where Aristotle says: 'It is not a body which has lost its soul which is such as to be potentially alive, but one which has [its soul]' (*esti de ou to apobeblêkos tên psuchên to dunamei on hôste zên, alla to echon*). The suggestion is that Aristotle denies that the non-organic body potentially constitutes the organic body, and indeed that he is right to do so given his application of the homonymy principle. But this would be to attribute to Aristotle the logical fallacy of which his detractors seem guilty, viz. that only what is already necessarily *F* can be considered as potentially *F*. This is false, and more important, is not entailed by the homonymy principle. Is Aristotle guilty of this same mistake? He is not, as the context of the passage makes clear. He goes on to point out that 'seeds and fruit are potentially bodies of this kind' (412ᵇ26–7). His point is evidently the mundane observation that one does not bring corpses back to life; one brings about new organic bodies from seeds. Thus, he evidently has in mind only *physical* necessity in this passage, and does not deny that it is conceptually possible that corpses could become, in the sense of come to constitute, organic bodies. To show Aristotle committed to the fallacy attributed to him, it is first of all necessary to show that he has in mind a form of necessity stronger than physical, but (a) the passage is compatible with weaker, physical necessity, and (b) the context suggests that this weaker sort is intended. Indeed, in this very passage Aristotle claims that certain sorts of non-organic bodies, viz. seeds, are in potentially organic bodies. With a univocal reading of potentiality, viz. physical, this entails that non-organic bodies are nomologically, metaphysically, and conceptually potentially organic bodies. Thus, not only does this passage not preclude the interpretation I have offered; it positively requires it.

might equally be the bronze in the statue of Theseus. But it is difficult to see how the organic body of Socrates could be anything other than the organic body of Socrates.[27]

The argument behind this objection would seem to be: (1) stuff x can be the matter of form F only if x and F have distinct modal properties; (2) given the homonymy principle, an organic body could not have modal properties distinct from its soul; (3) therefore, organic bodies cannot be the matter of souls.[28] One could of course grant the conclusion of this argument and say that the non-organic body has modal properties the soul lacks and that this is or should be what Aristotle regards as the matter of which the soul is the form. This would preserve Aristotle's analogy and governing insight and would moreover provide a clear way to understand his application of the homonymy principle to human bodies. But this account will not suffice, since (a) Aristotle quite clearly and self-consciously holds that the body which is the matter of the soul is the organic rather than the non-organic body,[29] and (b) it is moreover not clear that he should hold this view without several important qualifications.

---

[27] One might argue that although organic bodies are necessarily alive, this does not yet require that a particular body is necessarily the body of a particular soul and that, consequently, Socrates' organic body could in principle be informed by Callias's soul. If this were a bona fide possibility for Aristotle, then already we could point to analogies between organic bodies and more standard examples of Aristotelian matter. But I will overlook this possibility since (a) there is no evidence that Aristotle conceives of the relationship between particular bodies and souls in this way; (b) there *may* be some evidence that he actually rejects this picture in his discussion of the Pythagorean myths in *An.* i. 407^{b}20–4; and (c) even if this possibility were countenanced, it would remain to discuss the potentially deeper sources of disanalogy discussed in the text.

[28] There is another formulation of the conceptual argument which is in some ways closer to Ackrill's actual remarks. Cohen (1987, 118), for example, takes his point to be that 'on the one hand, the matter of any compound must *potentially* have that form; on the other hand, it must not have it *necessarily*' (italics as found). On this construal of the conceptual problem, the argument is simply: (1) x is the matter of F only if: (i) x is potentially F and (ii) x is not necessarily F; (2) organic bodies are not potentially ensouled; (3) organic bodies are necessarily ensouled; therefore, (4) organic bodies cannot be the matter of souls. On this reading, the case against organic bodies is overdetermined, at least if one follows Ackrill in holding that potentiality 'gives way' to actuality, so that what is actually F is not simultaneously potentially F. I have already given reason for believing that while (3) is true, Aristotle's distinction between organic and non-organic bodies renders it unproblematic (at least as regards generation). But I have chosen a more general formulation of the conceptual argument because I believe it will allow a more general treatment of considerations which could be taken to validate premise (1) in this more specific formulation.

[29] It might be objected that when Aristotle says 'organic' in this context, he does not intend the relatively complicated gloss of that notion I have given. He may intend, as some translators suggest, only that the body have organs (whether or not they are functioning), as opposed to inorganic bodies, which lack organs. If this were right, the discussion would be at an end, since if the non-organic body, which clearly persists through the death of the organism, were the proximate matter of the soul, then the body would not be necessarily alive and could have modal properties which the soul lacked. This cannot be Aristotle's

More important, one need not grant this argument at all. Premise (2) is first of all not established by noting either that souls and organic bodies exactly temporally overlap or that one depends upon the existence of the other for its existence (or even that *each* depends upon the other for its existence).[30] Indeed, Aristotle evidently regards souls and organic bodies as having distinct actual properties, and so trivially as having distinct modal properties, whether or not they temporally overlap.[31] Premise (1) is also without textual basis; indeed, some commentators will maintain that Aristotle positively rejects (1), arguing that diachronic material continuants, diachronic slices of proximate matter, are identical to particular forms.[32]

It might be objected that this formulation of the conceptual problem with the homonymy of the body fails to address the fact that even if organic bodies can in principle be the matter of souls, they are nevertheless disanalogous with paradigmatic cases of Aristotelian matter, given their necessarily perfect temporal overlap with the forms they instantiate. But this is debatable,[33] and a consideration of this objection will show

view, however, since in this instance, his application of the homonymy principle would be inexplicable. This shows in any case that the notion of an organic body is richer than merely 'a body with organs'. See *An.* 412^{b}1–4, where the simple organs of plants are discussed. It is a difficult matter to determine precisely what organic bodies are, but in the context of *An.* ii. 1, Aristotle evidently intends, roughly, 'a sort of body which is so constituted that it realizes the functional states definitive of human beings'. In this sense, as OB already indicates, it is merely analytic that organic bodies cease to exist when no longer ensouled.

[30] A useful parallel here would be Aristotle's analysis of the relationship between motion (*kinêsis*) and time (*chronos*) in *Physics* iv. 10 and 11. Although not reducible to motion (218^{b}18), time in some sense belongs to motion and is dependent on it (219^{a}1–2, 8–10, 251^{b}11–3). The matter is complicated, but it seems clear that Aristotle holds that (a) the existence of motion is necessary and sufficient for the existence of time (and therefore the existence of time is necessary and sufficient for the existence of motion), but even so (b) the existence of time depends upon the existence of motion in an asymmetric way. Similarly, organic bodies and souls will mutually entail the existence of one another, even if the particular psychic predicates of an agent supervene on physical predicates of that agent's body and could have been realized in other ways.

[31] I discuss these properties in Shields 1988*a*. In summary, they are: (1) the soul cannot be moved in itself (*kath' hauto*; *An.* i. 3), whereas every magnitude, including the organic body, can be moved *kath' hauto* (*Cael.* 268^{b}15–16); (2) the soul, as the form of the body, is not generable (*Met.* vii. 8, xii. 3), while the organic body is; (3) the soul is not divisible (*An.* 411^{b}27), while every magnitude is (*Phys.* 219^{a}11, 237^{a}11); and (4) the soul is neither one of the elements nor from the elements (*GC* 334^{a}10–11), while the organic body is.

[32] I myself reject this interpretation, in part for the reasons given in n. 31 above. The point I wish to make in the present context is that it is not obvious that anything Aristotle says constrains him to endorse this premise (we have already seen that nothing in his account of generation requires it).

[33] In this connection it is worth noting that when Ackrill suggests that the soul and body present special problems for Aristotle's hylomorphism, he neglects those passages in which Aristotle suggests quite generally that form and matter (at least proximate matter) are one, e.g. *Met.* 1045^{b}18–19: 'the proximate matter and the form are one and the same, the one in

how widely, and in what sense, Aristotle is prepared to regard material constituents generally as homonymous.

It is noteworthy that Aristotle extends the homonymy principle to artefacts, and in so doing demonstrates that there is nothing peculiar about organisms as regards the homonymy principle.[34] As we have seen, there is a clear sense in which organic bodies depend upon being enformed for their existence as organic bodies. But this is not a result of their being the bodies of living beings, but rather of Aristotle's functional analysis of kind membership. Artefacts of various sorts also fall under the homonymy principle, and indeed, artefacts of type F that lose the function characteristic of Fs are only discretely homonymous Fs. It is easiest to see why Aristotle is right to maintain this by considering artefacts whose functions are obvious. An 'axe' which cannot chop, Aristotle maintains, is a discretely homonymous axe (*An.* $412^{b}13$–15). This seems to entail that only matter in a certain condition counts as the matter of an axe. Matter which falls out of this condition is no longer the material of an axe; it is the material of something which has a superficial resemblance to an axe. Axes, like human beings, are functionally defined: all and only those things which can chop count as axes. Something will count as the matter for an axe, then, only when it is the matter of something which can chop. Consequently, 'axe matter', as we might call it, necessarily has the property of constituting an axe. When it ceases to constitute an axe, it ceases to be axe matter, even though it will remain potentially axe matter, that is, potentially constitutive of something which chops.[35]

---

potentiality, the other in actuality' (*hê eschatê hulê kai hê morphê tauto kai hen, <to men> dunamei, to de energeia(i)*). Minimally, these sorts of passages suggest that the identity conditions of the proximate matter are parasitic on the identity conditions of the particular form whose matter it is. Hence, if there is a problem about hylomorphism as regards soul and body, then the problem is really quite general. But I do not think there is such a problem.

[34] This is a point well made by Cohen (1987, 120–1) against Burnyeat. Although I agree with Cohen that Aristotle's extension of the homonymy principle to artefacts undercuts Burnyeat's use of it in his attempt to show that Aristotle's concept of matter is utterly alien to post-Cartesians, my account differs in relating this extension more explicitly to Aristotle's functionalism and his introduction of proximate matter. I might add that Aristotle's extension of homonymy to artefacts suggests there is something askew with Ackrill's question (1972–3/1978, 75): 'Is there something special about the concept of living things that makes it recalcitrant to Aristotle's treatment?'

[35] Williams (1985, 110) is somewhat misleading in saying: 'It is not that the dead eye, or any quantity of inanimate matter such as bronze or stone, *is* formless matter: no matter can exist apart from form. It is rather that inanimate body depends for its continued identity on its matter whereas the animate body does not.' Corpses, like other artefacts, can presumably admit of some material replenishment while remaining numerically one. I take Aristotle's point rather to be that functionally defined bits of matter remain members of the classes determined by those functions only so long as they retain those capacities.

Thus, Aristotle recognizes two ways of characterizing the matter which constitutes an axe, both functionally oriented: there is first the matter capable of realizing the form of an axe, e.g. iron, and second the matter - capable of chopping, e.g. axe-shaped iron. Of course, axe-shaped iron is capable of realizing the form of an axe, because it does, and thus the Lockean mass of iron which constitutes an axe at t_1 will be equally potential and actual axe matter. It does not follow that axe matter and the quantity of iron which constitutes the axe are identical; their identity conditions may differ even though they spatiotemporally overlap during the period in which we have a non-homonymous axe.

If this is so, then axe matter will necessarily have the property of being able to chop wood and will be directly analogous to organic bodies. An obvious difference may seem to be that we have axe matter when and only when we have something axe-shaped, while we can have something human-shaped which is not composed of organic matter, viz. a corpse.

But this difference does not point to an interesting disanalogy. It derives only from the fact that the function of certain sorts of artefacts is directly dependent, in part, upon their manifest shape. But when he speaks of the form (*eidos*) of an axe, Aristotle clearly has in mind not the outward shape as such, but rather the essence of what it is to be an axe (*to ti ên einai*), which in turn requires a certain outward shape. This shape may be necessary for something's being an axe, but it is surely not sufficient. Given his commitment to FD, Aristotle rightly thinks of certain sorts of artefacts as discretely homonymous *F*s, even though they may have precisely the outward shape of genuine *F*s. The wooden saws introduced at *Meteorologica* 390ᵃ10–15, for example, may bear a superficial resemblances to real saws, even though they lack the capacities requisite for being actual saws. The problem with wooden saws is not that they have the wrong shape, but rather that *in virtue of their matter* they cannot perform certain functions. Their matter is incapable of realizing a certain form, functionally construed, even though it can take on the superficial shape which normally accompanies that form. Thus, Aristotle clearly holds that the range of matter suitable for realizing a given form will always be functionally constrained: only matter of a certain sort or complexity will be potentially the matter for a given form.

In this sense, any structured type of matter will be precluded from realizing a given range of forms.[36] Although an appropriate stuff for statues, bronze will never realize the form of being a ham sandwich—even though

---

[36] By structured matter I mean any matter at or above the elemental level. I do not take a stand on whether Aristotle accepted the existence of prime matter.

in the Museum of Modern Art we could find a discretely homonymous brazen ham sandwich. It is fair to suppose, then, that Aristotle holds that only materials capable of supporting certain structures will be potentially members of a given functionally delimited class. But if so, we should not be surprised to find him applying the homonymy principle indifferently to organisms and artefacts. Indeed, this is precisely what we should expect given FD.

The symmetry with which the homonymy principle functions as regards artefacts and organic entities can perhaps best be illustrated by considering an artefact not available to Aristotle, but one whose introduction will do no violence to the issue under consideration. If we consider two qualitatively identical personal computers, one of which we unplug, and one of which we do not, we can see how neatly the homonymy principle applies to artefacts. We might think that something is a computer only if it is able to realize certain sorts of functional states. There is a sense, of course, in which the computer without electrical power cannot realize these states, and it is precisely in this sense that it is a discretely homonymous computer. A computer in a world without electricity would not really be a computer at all, just as a body without life is not really a human body at all: both have every outward and structural appearance of their genuine counterparts, but in virtue of their lacking determinative capacities, neither is itself a genuine member of those kinds.

When considering a similar case, Williams suggests, 'You have to be very resolutely Aristotelian to deny that a stopped clock, even a clock that will never start again, is a clock, unless it is in advanced physical dilapidation.' [37] Perhaps, but one would surely be right to do so if the clock will not stop because it loses its capacity to tell time. We, of course, customarily offer the judgement that a particular artefact is a clock on the basis of its manifest structure, and for this reason we might initially regard a discretely homonymous clock as a clock. But just as we were right to count digital clocks as clocks, even though they bear little structural or compositional resemblance to analogue clocks, so we would be wrong to maintain that permanently stopped clocks are genuine clocks, rather than as seductive, discrete homonyms. Of course, stopped clocks and powerless computers are, in our world, readily altered so as to become genuine clocks and computers, whereas corpses are not.

Here it is worth noting that our modal intuitions are governed in part by the ways in which an entity is precluded from fulfilling its function. Something may lose its intrinsic capacities, or it may retain those capaci-

---

[37] Williams 1986, 189.

ties but be prevented from actualizing them because of the loss of some external enabling conditions. Thus, we will be inclined to regard an unplugged computer as a computer; we may feel somewhat inclined to regard something structurally equivalent to a computer in a world without electricity or any other possible source of energy as a non-computer, or we may focus on its retention of its internal structural capacities and insist that it remains a computer; yet we should not regard something which appears to be a computer but lacks any internal components as a computer at all. Aristotle's functional determination thesis is best understood as ranging over the *intrinsic* capacities of an entity. This is presumably what is meant when he claims, for example, that a capacity is a 'source (*archê*) of change in another thing or <in oneself> *qua* another' (*Met.* 1046^{b}10–11; cf. 1019^{a}15–16, *Phys.* 251^{a}11, ^{b}1–5). In calling a capacity an *archê* in this connection, Aristotle treats in as an internal source of motion, one which features into an entity's structure, rather than as a necessary background condition. Consequently, only when an intrinsic capacity for performing the activities determinative of *F* is lost does an entity cease to be an *F*. Stopped clocks do not lose such capacities. Hence, we need not be made uncomfortable by either Aristotle's functional determination thesis or by the consequences regarding discrete homonymy which he rightly derives from it.

Hence, worries about non-functioning artefacts should not incline us to question the tenability of Aristotle's conception of the body's homonymy. There would a minor problem here for Aristotle only if he failed to distinguish between internal capacities and the circumstances which permit those capacities to be realized; but he makes just this distinction (*Met.* 1048^{a}13–24).

Consequently, and in summary, the homonymy principle does not present any deep problem for Aristotle's hylomorphic analysis of body and soul. The homonymy principle would result in problems for Aristotle's account of generation and destruction in terms of the acquisition or loss of substantial forms only if he held to unwarrantedly strong requirements regarding the persisting *hupokeimena* in these changes. But he does not; on the contrary, he rejects such strong constraints. Nor does the homonymy principle land Aristotle in any general conceptual problem within his hylomorphic scheme. First, the general conceptual argument invokes premises which are both without textual basis and in any case independently dubious or false. Second, bodies are not peculiar in being subject to the homonymy principle: Aristotle reasonably (given his commitment to FD) extends it to artefacts as well. Hence, Aristotle's doctrine of the homonymy of the body remains unproblematic within the framework of hylomorphism.

5. 6 THE BODY'S HOMONYMY AND
FUNCTIONALISM

Some have suggested that beyond the general problems already discussed, the homonymy principle presents special difficulties for functionalist interpretations of Aristotle's philosophy of mind.[38] If we take seriously Aristotle's commitment to FD, his functional analysis of kind individuation, then we have reason to regard him as a thoroughgoing functionalist, and I will now argue that the homonymy principle presents no difficulties at all for this view. On the contrary, the homonymy principle is compatible with functionalism in Aristotle's philosophy of mind, and indeed both spring equally from Aristotle's commitment to the functional determination thesis.

The homonymy principle gives rise to two potential problems for functionalist interpretations of Aristotle, in view of: (1) functionalism's commitment to there being nothing more than a contingent connection between functional and material states, and (2) functionalism's well-motivated commitment to the multiple realizability of the mental. These points are related, since an account which shows (2) to be compatible with functionalism will equally serve to do so for (1).

The second problem might be explicated as follows. Functional states are multiply realizable; any material (or conceivably immaterial) state which realizes a given functional state will equally be in the psychic state identified with that state. The appropriate bases for functional realization are not constrained by functionalism *per se*, and thus *anything* capable of realizing a given functional state will equally be capable of realizing the associated mental state. But this seems incompatible with the homonymy of the body: in applying the homonymy principle Aristotle implicitly restricts the range of subjects which have the potential for realizing psychic states. Indeed, this range is radically restricted. The only possible candidates are organic bodies, which are necessarily alive. If this is so, then the homonymy principle entails that Aristotelian mental states are not multiply realizable and therefore cannot be identified with functional states. In short, the homonymy of the body precludes the commitment to the multiple realizability of the mental required by functionalism.

This argument shares a defect with the general argument already considered. The homonymy principle does not constrain Aristotle to hold that the only *potential* bases for realizing mental states is the organic body. He does think that in fact only organic bodies realize human psychic

---

[38] Burnyeat 1992.

states, but this does not entail that other sorts of bodies are not potentially organic. As we have seen, non-organic bodies are potentially organic; they cannot be identical with organic bodies, but this does not preclude their coming to constitute them. If this is correct, it is false that the only potential bases for realizing mental states will be the organic body. Hence, the multiple realizability of the mental is compatible with the homonymy principle. Therefore, the homonymy principle poses no threat to a functionalist reading of Aristotle's philosophy of mind.[39]

On the contrary, there is an interesting conceptual connection between the homonymy principle and Aristotle's functionalism. Each is entailed by a deeper metaphysical commitment of Aristotle's, viz. the functional determination thesis. In this sense, Aristotle's views cohere: the homonymy of the body, together with functionalism in the philosophy of mind, form a consistent and defensible set of views about mental states and their realizations.

---

[39] Indeed, if Aristotle regarded the homonymy principle as a hindrance to the multiple realizability of human beings, we should be surprised to hear him say, when determining 'whether it is necessary that the account (*logos*) of the parts . . . be present in the *logos* of the whole' (*Met.* 1034^{b}22–4), that although we always see human beings realized in flesh and blood, 'nothing hinders' their being realized in other stuffs. He compares human beings with brazen circles, suggesting that if all circles were realized in bronze, as all human beings are realized in flesh, we would have trouble realizing that circles could be realized in any functionally suitable matter. Williams (1986), who thinks the homonymy principle has unacceptable consequences for Aristotle, comments on this passage: 'There is a passage—*Metaph.* 1036^{b}1 f.—which might be taken to mean that the human *form* could be realized in something other than flesh and bones, etc. But I doubt that this is what he means: Aristotle thinks it necessary that they go together, but that we can (though for this very reason we find it difficult) abstract the form from the matter' (italics as found; 192 n. 2). Williams's account of this passage is perplexing: if in the end we can abstract the form from the matter, then it is unclear how it is 'necessary that they go together'. To make Williams consistent, we need to read 'necessity' as nomological necessity or as necessary coextension. But if Aristotle thinks only that matter and form are nomologically related, then his views are compatible with the sort of multiple realizability required by functionalism; if he thinks that they are merely necessarily coextensive, then Aristotle should also allow that they are separable in account (*choriston logô(i)*), in which case nothing about the account of form will preclude realizability in different functionally suitable types of matter. Hence, Williams's interpretation of this passage either is internally inconsistent or fails to show that what Aristotle says is incompatible with the multiple realizability of the human form. In either case, this passage provides further support both for my analysis of the homonymy principle and for the attribution of functionalism to Aristotle. Williams does not comment on an earlier paper dealing with this passage (Heinaman 1979, 261 n. 20), in which Heinaman observes: 'Aristotle's position as stated is not that it is unclear whether the matter is not part of the form, but rather that if the form always appears in a certain kind of matter, unlike the form of a circle, then abstracting the form is more difficult. That is, the difficulty is epistemological, not metaphysical (1036^{b}2–3, 7).'

5. 7 THE RELATION BETWEEN ORGANIC AND NON-ORGANIC BODIES

Aristotle's distinction between organic and non-organic bodies seems both well motivated and serviceable in analysing hylomorphism in light of the demands of his commitment to homonymy. But by availing himself of this distinction, Aristotle introduces an additional set of players into what seems an already complicated set of *relata*: souls (*psuchai*), compounds (*ta ex tôn amphoin*), and now both organic bodies (*sômata organika*) and non-organic bodies (*sômata*).

Now, organic and non-organic bodies are clearly not identical, since they have diverse modal and historical properties (non-organic bodies exist when organic bodies do not). But whatever their connection, it would seem to be rather intimate. It will not do to understand the relationship between organic and non-organic bodies as one of temporal succession; non-organic bodies do not come into existence only when organic bodies go out of existence. Corpses are a phase of the existence of non-organic bodies, but non-organic bodies exist before and survive the death of the organism. Just as the iron of an axe co-exists with axe matter, so the non-organic body exists while organic bodies exist.

The relationship Aristotle requires between organic and non-organic bodies is not at all mysterious and, as I have already hinted, is fruitfully understood as some form of constitution. As entities which exactly overlap spatiotemporally so long as the organic body exists, organic and non-organic bodies will have all their non-modal, non-intensional properties in common. And this is sufficient for constitution.[40]

This is the sort of relationship Aristotle evidently has in mind, in any case, when distinguishing between grades of proximate and non-proximate matter in terms of their sophistication. He complains that clay, for example, is not appropriately regarded as the matter of a house, on the grounds that such matter is too far removed from the actual materials used by the builders (*Met.* 1044^b1–3, cf. 1049^a21–4). Rather, it is the bricks compounded from clay which is appropriately regarded as the proximate matter (*eschatê hulê*) of the house. Of course, he recognizes that it will be in one way true to say that clay is the matter of the house, precisely because the bricks are composed exclusively of clay; but this will be true only

---

[40] Constitution in this context is simply meant to be a form of oneness weaker than identity: x constitutes y if and only if x and y overlap in space and time, have all of their non-modal properties in common, and though there is no difference in y without a difference in x, y can persist through material replenishment. Thus, so much clay and paint will constitute a Ming vase; the clay and paint may have little worth when they are no longer vase-shaped, but so long as they are, they have just the value of the vase.

derivatively. Only by constituting the bricks, the proximate matter, does the clay have any claim as a constituent of the house.

Much the same can be said about the organic and non-organic bodies. Given his introduction of proximate and non-proximate matter, Aristotle has at his disposal a clear account of the relationship between organic and non-organic bodies. Non-organic bodies, like the clay of which bricks are compounded, constitute organic bodies, which in turn serve as the proximate matter of individual souls. But the further relationship between a particular organic body and an individual soul, as that between so many bricks and the form of a certain house, is not given by the relationship between the grades of proximate and non-proximate matter and must be settled on independent grounds.

Also to be settled on independent grounds is the question of how Aristotle can regard the organic and non-organic bodies as both one and not one. For we might typically come to regard the matter which constitutes something, clay for example, and that which it constitutes, a Ming vase, as one thing. Certainly, for example, we would be disinclined to buy the clay after already having paid for the vase. Since he is willing to distinguish not only the vase and the clay which constitutes it but also the proximate and non-proximate matter, Aristotle will need to specify the forms of oneness and sameness involved more precisely than he has thus far.[41]

5. 8 CONCLUSIONS

Aristotle regards certain bodies, including corpses, as discrete homonyms. He is not, however, in a position to withdraw his application of the homonymy principle to bodies. In so far as the homonymy principle springs from Aristotle's commitment to a functional analysis of kind membership and individuation, he is constrained to regard corpses as akin to statues and paintings of bodies—as discrete homonyms. Nor should he abandon it: the homonymy principle does not lead Aristotle into inconsistencies or theoretical problems of any kind, applied either to human bodies or to artefacts generally. Further, the homonymy principle presents no difficulties for functionalist interpretations of Aristotle's philosophy of mind. On the contrary, the principle of the homonymy of the body and a weak variety of functionalism are equally entailed by Aristotle's functional determination thesis. Aristotle's views on these matters form a coherent and defensible package.

---

[41] I return to this issue at 6. 7.

That said, it is equally true that no one, in the course of daily affairs, would have recourse to a distinction between organic and non-organic bodies. This is because ordinary circumstances do not call for any such distinction. This, then, is why some discrete homonyms may escape our notice: the problem is that we fail to note that applications of F in 'a is F' and 'b is F' have different accounts, because we never have occasion to notice that they might. When an uncommon context provides reason for questioning whether the Fs in question may have distinct accounts, we may be reluctant to suppose that they do. The only justification for coming to believe that they do will result from the application of an independently motivated principle, like FD, which proves at odds with customary usage. If these broader principles are motivated, then so too will be Aristotle's appeals to *mere* homonyms, or seductively discrete homonyms.

Even so, few of Aristotle's appeals to homonymy in serious philosophical contexts are appeals to discrete homonymy. For it remains true that most discrete homonyms are patently non-univocal. Consequently, their non-univocity rarely requires special pleading.

The same is not true of associated homonymy, including especially core-dependent homonymy. I turn now to some putative instances of associated homonymy, the first of which helps explain how organic and non-organic bodies can both be and not be one.

6

Oneness, Sameness, and Referential Opacity

6.1 FORMS OF ONENESS AND SAMENESS

Aristotle's introduction of two bodies, one organic and one non-organic, may seem in various way ad hoc; it may also seem unparsimonious in the extreme. After all, the organic body occupies the same space as the non-organic body for its entire career, although the non-organic normally persists for a while after the demise of the organic body. It may further appear to be introduced merely for the purpose of staving off telling objections to hylomorphism. If this distinction has no independent motivation, then critics will be right to find Aristotle's appeals to two spatiotemporally overlapping bodies unappealing. They will also be right to wonder whether the sorts of principles adduced to motivate a distinction between organic and non-organic bodies might not multiply entities beyond control.

Aristotle ought to be sensitive to these sorts of worries. For he himself derides the Sophists for their treatment of accidental oneness in ways which suggest an awareness of allied problems. The Sophists want to know, when Coriscus is both a musician and adept at grammar, whether the musical thing (*to mousikon*) and the grammatical thing (*to grammatikon*) are the same or different; they also want to know whether the musical Coriscus and Coriscus are the same (*Met.* 1026^{b}15–18). Aristotle thinks Plato is partially justified in accusing them of trading in non-being (*Met.* 1026^{b}14–15; cf. *Phys.* 191^{b}13–17), since 'the accidental seems to be something near non-being' (*Met.* 1026^{b}21). Aristotle's complaint here cannot be that the Sophists are wrong to take up the question of accidental unity as such: he considers it himself, at length. Rather, they make a muddle by exploiting the phenomenon of accidental unity for the contrivance of paradox (*Met.* 1026^{b}18–21). The philosopher, then, needs an account of accidental unity himself, if only to circumvent the paradoxes others derive from it.

Aristotle's anti-Sophistic justification for inquiring into accidental unity is wholly defensible. Yet his interest in the topic extends far beyond what such a defensive posture warrants. Indeed, Aristotle points to other reasons why philosophers should engage questions of accidental unity. Most generally, it falls to them in so far as it is their function to theorize about all things: 'For if not the philosopher,' he asks, 'who will it be who inquires into whether Socrates and Socrates seated are one and the same?' (*Met.* 1004ᵇ1–3). More specifically, it falls to philosophers to determine how accidental unities function, if at all, (i) in generation and destruction (*Phys.* i. 7); (ii) in questions of the nature of identity and other forms of oneness (*Met.* vi. 6, 9); and, most strikingly, (iii) in questions of substitutivity *salva veritate* in opaque contexts, where co-referential terms are inter-substituted without preserving truth (*Top.* i. 7; *Soph. El.* 179ᵃ1–10).

In this chapter I investigate Aristotle's conception of accidental sameness within the framework of homonymy already developed. One advantage of proceeding in this way is that we will be able to avoid attributing to Aristotle a peculiar ontology of what commentators have variously called 'kooky objects',[1] or 'modally individuated objects',[2] or what I will call 'hyper-finely-individuated objects'. This is an advantage in so far as it spares Aristotle the only dubiously coherent notion of relative identity.[3] It is also an advantage because it provides a way to respect Aristotle's clear commitment to a privileged ontology of intrinsic unities, a commitment proponents of kooky objects are compelled to ignore.

I begin by considering Aristotle's analysis of change and generation in *Physics* i. 7, since some have wanted to find his remarks there congenial to an ontology of hyper-finely-individuated objects. This discussion also serves to introduce the principal issues surrounding an ontology countenancing such objects and so helps show what is at stake in determining whether Aristotle accepts any such view.

6. 2 ACCIDENTAL ONENESS IN ARISTOTLE'S GENERAL THEORY OF CHANGE AND GENERATION

In *Physics* i. 7 Aristotle introduces the notions of form and matter in an effort to analyse change, broadly construed to include both qualitative change and generation. With an eye on some Eleatic puzzles about the very possibility of change and generation,[4] Aristotle analyses all change

---

[1] Matthews 1982.

[2] White 1971 and Code 1976.

[3] On relative identity, see Wiggins 1967 and 1980; Hirsch 1982; and Yablo 1987.

[4] For discussions of Aristotle's analysis of change against the Eleatic challenge, see Wieland 1962, 112; Charlton 1970; Bostock 1982; Waterlow 1982; Code 1976, 360; and Graham 1987, 133–52.

as involving the persistence of some underlying continuant, and the acquisition or loss of some form, or positive quality.[5] For example, if Coriscus becomes musical, then we may say that Coriscus was a non-musical man who became musical. Coriscus, the man, persists through the change, having gained a positive attribute he previously lacked.

Aristotle seeks, with some difficulty,[6] to extend this account to substantial generation:

> As coming to be is spoken of in many ways, in some cases there is not coming to be <simpliciter> but a coming to be some particular thing, whereas among substances alone is there a coming to be simpliciter. And in all other cases <than substantial generation> it is clear that it is necessary that something underlie what comes to be, for when something comes to be a certain quantity or a certain quality or in relation to something else or in a certain time or place, <there is> something underlying, because only substance is not said of something else as underlying it, and everything else is said of substance. That substances too, and whatever else comes to be simpliciter,[7] comes to be from something underlying would be clear to one examining <the matter>. For there is always something which underlies, from which what comes to be <comes to be>, for example plants and animals from a seed. (*Phys.* 190ᵃ31–ᵇ5)

On the basis of this analysis of change, Aristotle infers that 'everything which comes to be is always a complex' (*to gignomenon hapan aei suntheton esti*; *Phys.* 190ᵇ11).

Aristotle's analysis gives rise to a host of problems.[8] These include a problem about the ontological status of the complexes (*suntheta*) which come and go. Even in the simple case of non-substantial change, Aristotle submits that we can conceive of someone's becoming musical in three ways (*Phys.* 189ᵇ34–190ᵃ2):

(1) The man becomes musical (*gignesthai anthrôpon mousikon*).

(2) The unmusical thing becomes musical (*to mê mousikon gignesthai mousikon*).

(3) The unmusical man becomes a musical man (*ton mê mousikon anthrôpon anthrôpon mousikon*).

Aristotle distinguishes between simple (*hapla*) and compounded (*sugkeimena*) cases: (1) and (2) are simple cases, since both the things which

---

[5] Here *eidos* is introduced in a quite general way, without the more technical function it acquires only later in Aristotle's developing hylomorphism.

[6] On which see Ross 1936, 492.

[7] I accept Ross's rejection of *alla* at *Phys.* 190ᵇ2. See Ross 1936, 492.

[8] For fuller discussions of the problems of substantial generation and unity to which these principles give rise, see Gill 1989 and 1994; Dancy 1978; Irwin 1988, 237–8; Lewis 1982 and 1994; Charles 1994; Charlton 1994; Scaltsas 1994; Fine 1994; and Spellman 1995.

become something (men and unmusical things) and the things they become (musical) are not compounded out of several things, whereas in (3) both what becomes (the unmusical man) and what it becomes (a musical man) are compounded.

Here Aristotle's account may take on a perplexing hue. From which simple things (*hapla*) are the compounded things (*sugkeimena*) compounded? It seems that there are two simple things in case (3) and that on Aristotle's approach the unmusical thing and the man are compounded into a musical man. Although this may seem perverse, Aristotle lends some credence to this way of thinking when concluding, rather disarmingly: 'The one thing which comes to be survives, while the other does not survive; what is not an opposite survives (for the man survives), but unmusical thing or the nonmusical thing does not survive,[9] nor does what is compounded from both of them (*oude to ex amphoin sugkeimenon*), e. g. the non-musical man' (*Phys.* 190ª17–21; cf. 190ª8–13). The unmusical thing does not survive when the musical thing comes to be. So, too, the non-musical man perishes when the musical man comes to be.

According to one contemporary interpreter, 'The implications of this doctrine are staggering.'[10] Why are they staggering? It is hardly problematical that we can secure reference to entities described in complex ways. If I refer to 'the puzzled redhead in the corner', I may pick some individual for discussion, perhaps the same individual you refer to as 'the stout little worrier'. In these cases we demonstrate the same entity under different descriptions and present it under different perceptual modes of presentation. If the puzzled redhead is the stout little worrier, our communication about him can proceed effortlessly.

Our communication proceeds effortlessly *if* we designate the same entity in different ways and have mutual knowledge that we have done so. We could go awry in several ways, for example, if the stout little worrier you indicate happens to be a puzzled redhead sitting in a corner other than the one in which my puzzled redhead is sitting. This is still no problem and is hardly staggering; it is, rather, a simple case of miscommunication. Yet

---

[9] In translating *to mê mousikon de kai to amouson ouch hupomenei* at *Phys.* 190ª19–20 as 'unmusical thing or the nonmusical thing does not survive', I presume that Aristotle uses the neuter expressions as substantives. It is possible, though unlikely, that Aristotle uses these locutions non-substantively ('what is unmusical' or, periphrastically, 'the trait of being unmusical'). This is highly improbable, given the co-ordination of *to anthrôpô(i) einai* and *to amousô(i) einai* at *Phys.* 190ª17. The contrast which follows ('what is compounded from both of them') is intended not to distinguish compounds from attributes, but rather to distinguish compounds from their discernible constituents. In any case, if these phrases were understood non-substantively, the case of 'kooky objects' would be still less compelling.

[10] Matthews 1982.

at least on some interpretations, Aristotle's doctrine may suggest another, altogether unexpected way we could go awry: even though the puzzled redhead is in the same corner as the stout little worrier, is sitting in the same chair as him, and indeed occupies precisely the same place in all respects, we are nevertheless talking about two different entities. What you demonstrate when you use a demonstrative ('*That* one there?') is not the same thing I demonstrate ('Yes, that one there'). I am talking about the puzzled redhead, and you are talking about the stout little worrier, and *these are not the same.*

This would be rather staggering. The results of individuating objects so finely yields some immediately counter-intuitive results. When a seated woman stands, the seated woman ceases to exist and a new woman comes into existence, namely, the standing woman. When a child yawns, a fleeting, yawning child comes into existence, then perishes, and then a (new?) non-yawning child is born. Admitting relational changes of the sort Aristotle himself mentions (*pros heteron* at *Phys.* 190ª35), the matter becomes still more complicated. When Sally sneezes near Larry, a man-in-the-room-where-Sally-sneezes comes into existence and then perishes in an instant. The man with this short life span is not Larry, for Larry has a long life span and both pre-exists and post-dates the man in question.

The worry is that Aristotle's theory of change and generation commits us to analysing cases of alteration by appealing to a plethora of entities whose existence we had failed to mark. Now, there is nothing distinctive in analysing (3), 'The unmusical man becomes a musical man', as:

$$(\exists x)(\exists y)((Fx \ \& \sim Gx) \text{ at } t_1) \ \& \ (Fy \ \& \ Gy \text{ at } t_2)),$$

so long as we add implicitly or explicitly that:

$$(y = x).$$

The peculiarity some have located in Aristotle's theory of change is precisely that it precludes our adding '$(y = x)$'. Rather, Aristotle's approach is alleged to entail:

$$(\exists x)(\exists y)[((Fx \ \& \sim Gx \text{ at } t_1) \ \& \ (Fy \ \& \ Gy \text{ at } t_2)) \ \& \ (x \neq y)].$$

If Aristotle's account does have this entailment, it will have a host of arresting and peculiar features.

On the present understanding of Aristotle's theory of change, he commits himself to a doctrine of hyper-finely-individuated objects, which as a first approximation is just this:

> HFIO: For all non-identical properties ϕ and ψ, then if ϕx and ψy, x and y are not identical.

Any property available for securing reference is sufficient for individuating an entity; any two non-identical properties individuate entities such that they are non-identical.[11] Thus, if Coriscus is a musician who is seated at the moment, there is a musician and a seated man, and the seated man is not identical with the musician. Of course, if Coriscus is a musician who is seated at the moment, then there is a musician and there is a seated thing. Why should Aristotle's theory of change be thought to entail that they are non-identical?

As we have seen, merely referring to Coriscus variously as 'the musical thing' and the 'seated thing' in no way requires us to treat the things as other than identical. Nothing about the properties themselves requires non-identity. These properties are compatible and may easily be instantiated by the same object. (By contrast, if we knew that x was even and y odd, we would already know that x and y were non-identical.) Perhaps the thought is that since he says that the non-musical thing perishes whereas the man does not, Aristotle must be thinking of them as non-identical. After all, two things are identical only if they share all their properties, and we have just identified a property the musical thing has but Coriscus lacks, namely, perishing at t_1.

Now, nothing about Aristotle's bare locution 'the non-musical thing perishes' provides license for this sort of application of Leibniz's Law. I may say, in the appropriate context, 'The president of the Supreme Soviet no longer exists', and what I say will be true even though the former president of the dissolved Supreme Soviet continues to exist. For there is someone, Gorbachev, who is identical with the former president.[12]

Perhaps there is some stronger reason, peculiar to Aristotle's conception of change, for holding him to HFIO. Some features of his language may suggest such a reason. As we have seen, he says 'everything which comes to be is always a complex' (*to gignomenon hapan aei suntheton esti*; *Phys.* 190ᵇ11). Moreover, he refers to the musical man as 'compounded' (*sugkeimenon*), indeed as compounded from two things (*ex amphoin*). Perhaps the thought is that complexes and compounds are complexes of

---

[11] Some qualifications moderating this principle are, of course, possible. The first, most obvious suggestion would be that ϕ and ψ cannot entail one another, since then Aristotle will end up holding e.g. that this man is not the same as this animal. The proponents of kooky objects or modally individuated objects have not added this qualification but may be entitled to presume it. If they do not, then they will hold a maximally hyper version of HFIO: the duck swimming alone in the lake will not be identical with the animal swimming alone in the lake, the woman with two feet will not be the same as the woman with at least one foot, and so forth. I presume that the proponents of HFIO will insist on this much restraint. If they do not, kooky objects are kookier still.

[12] Wiggins (1967 and 1980) aptly diagnoses the kind of error committed here.

discrete entities and so are compounded out of them. This would provide some ground for ascribing HFIO to Aristotle, if it could be shown that he held to a mereology according to which all compounds and complexes were compounds of discrete, non-identical entities.[13]

No evidence suggests that Aristotle holds any such view. First, nothing in the theory of change adumbrated in *Physics* i. 7 presupposes any such mereology. Hence, someone wanting to find HFIO in that theory would need to augment it with an external commitment on Aristotle's part. For this reason alone, it is wrong to understand Aristotle's hylomorphic analysis of change as requiring HFIO.[14] Moreover, it is unlikely that other passages state or even suggest the sort of mereology required to infer HFIO from Aristotle's account of change. On the contrary, Aristotle's anti-Sophistic rhetoric betrays a defensible hostility to HFIO.

6.3 HFIO: THE ISSUES

The reaction of some to Aristotle's discussion of change and generation in *Physics* i. 7 helps sharpen the issues surrounding HFIO. Some things he says are clearly not problematical. First, it is not problematical for Aristotle, or anyone else, to use singular terms that pick out one entity under distinct descriptions.[15] He may refer to Socrates as 'Socrates' or as 'the seated one' or as 'the walker' indifferently. It is, moreover, not problematical for Aristotle to talk of what is designated in these ways quasi-substantively. That is, he may freely talk of the seated one as the seated *thing* or as the seated *entity*.[16] Here it is possible to stipulate in a neutral, deflationary way that whatever is picked out by a singular term is some thing or other, without any categorial implications whatsoever.

Problems arise only when these relatively bland claims are strengthened and augmented by more distinctive theses. For example, if hyper-finely-individuated objects were held to be substances, in Aristotle's sense, then there would be a problem, since they quite evidently do not satisfy Aristotle's set criteria for substantiality. For instance, the seated thing

[13] The Revised Oxford Translation (Barnes 1984) is somewhat misleading here: 'One part survives, the other does not: what is not an opposite survives (for the man survives), but not-musical or unmusical does not survive, nor does the compound of the two, namely the unmusical man' (*Phys.* 190ª17–21). When Aristotle claims '*to men hupomenei, to d'ouch hupomenei*' (190ª17–18), he does not speak in terms of *moria*, in the sense of *proper* part.

[14] Matthews (1982) is therefore not justified in inferring that Aristotle holds a doctrine of kooky objects from this passage.

[15] For referring via modes of appearance, see 3. 6.

[16] On the status of Aristotle's habitual use of neuter adjectives in these sorts of contexts, see n. 9.

asymmetrically depends on the man for its existence, and so does not satisfy Aristotle's demand that substances be separate, where this involves minimally a capacity for independent existence.[17] Still, proponents of HFIO have not typically introduced hyper-finely-individuated objects as Aristotelian substances. On the contrary, they have mainly wanted to deny any such claim. Hence, no immediate categorial problems result from HFIO taken by itself.

We recognize a more serious problem when we see how HFIO treats the relation between the objects picked out under diverse descriptions. We have granted that it is unproblematical that it is possible to pick out entities under all manner of singular terms. Exceptional about HFIO is its claim that whenever we pick out entities under distinct definite descriptions, we pick out distinct entities. That is, whenever we refer to something as 'the ϕ' and then refer to something as 'the ψ', we know in advance that the things we have picked out are not the same. Thus, however prepared we might otherwise be to allow that 'Socrates' picks out something identical with the thing picked out by 'the seated man', we will have to concede, given HFIO, that they are not the same thing but are two things coinciding in space and time.

This much may already seem counter-intuitive. Even so, HFIO could none the less be justified by the philosophical work it might do in some larger context, for example, in Aristotle's postulation of organic and non-organic bodies, in his metaphysics more generally, or in his treatment of problem cases of opacity. The same cannot be said, however, when we reflect on some additional costs of HFIO, costs which should make us still less inclined to ascribe it to Aristotle. We see these costs by differentiating between stronger and weaker formulations of HFIO, differentiations not often marked by its proponents.

In its current formulation, HFIO entails that no two things are ever strictly identical. Wherever there are two properties, there are two things: if the seated man is also a musical man, then there are two men, and these are not identical. Further, entities normally thought to be co-extensive will be distinct. Creatures with hearts will not be the same as creatures with kidneys, since 'the one with the kidney' will not pick out the same individual as 'the one with the heart' in any given context. Of course, creatures with hearts may coincide with creatures with kidneys; but they cannot be identical with them. What is worse, without some restriction on HFIO, even necessarily co-extensive entities will be distinct. Since triangularity is not the same property as trilaterality, trilaterals will not be identical with triangles. Perhaps if they *necessarily* coincide, according to the propon-

---

[17] On the form of separation required for substantiality, see Fine 1983.

ents of HFIO, they will then be *necessarily accidentally the same*. This would be an odd position for Aristotle to embrace. Perhaps it seems to some that Aristotle's account of accidental oneness requires him nevertheless to embrace it.

6.4 ACCIDENTAL ONENESS AND HFIO

It is, of course, possible that Aristotle both criticizes the Sophists for their alleged adherence to HFIO and then comes to adopt some version of it himself, perhaps because he comes to see that the Sophists were correct after all or perhaps because he comes to embrace a version of HFIO without the consequences the Sophists must accept for their version.[18] Aristotle might come to agree with the Sophists for a number of reasons. If he has no concept of identity available to him, Aristotle might trivially be thought to hold that no two objects are identical. If Aristotle allows that both 'Socrates' and 'Socrates seated' are singular terms referring to objects in the world, and he has no concept of identity, then he will never have opportunity to claim that the objects referred to are identical; nor will he, by the same token, have opportunity to affirm that they are non-identical.[19] Still, as some have suggested, other features of his account may implicitly commit him to their non-identity.[20]

Less radically, Aristotle may have the concept of identity but also recognize weaker forms of oneness and unity.[21] If he recognizes weaker forms of oneness, Aristotle is free to exploit them in his analysis of the relation between Socrates and Socrates seated. Some features of his account of predication, for example, might incline him to do so.

---

[18] Matthews (1982) seems to suggest that Aristotle changes his mind about HFIO, but he does not speculate about what precipitated the change.

[19] White (1971) does not seem to appreciate this point.

[20] White (1971) argues this way; he is rightly criticized by Miller (1973).

[21] Spellman (1995) understands Aristotle as distinguishing between numerical sameness and identity. Socrates and Socrates in so far as he is a specimen of a natural kind are numerically the same, but not identical (30–1), though the notion of numerical sameness she ascribes to Aristotle is something she also regards as only dubiously coherent (100). Two observations: (i) the problems introduced by HFIO are perfectly general and so pertain to non-natural as well as natural kinds; hence, even if Aristotle means to distinguish identity and numerical sameness by something's being a specimen of a natural kind, this will not show how the notion of numerical sameness short of identity helps explain the general phenomenon of accidental unity; and (ii) the appeal to Leibniz's Law necessary to ground this distinction in the first place (not everything true of Socrates, the composite substance, is true of Socrates, the specimen of the natural kind human being, e.g. Socrates is snub-nosed, whereas the specimen of the natural kind, who has only those properties contained in the definition of the species, together with any those might entail, is not) is unconvincing.

I will argue that Aristotle has a concept of identity and that he recognizes forms of oneness weaker than identity. Even so, his motivations for recognizing forms of oneness weaker than identity provide no evidence that he accepts HFIO.

6. 5 HFIO AND THE FALLACY OF ACCIDENT

We have seen so far that nothing in Aristotle's theory of generation or accidental unity commits him to HFIO. If he needed to appeal to HFIO in either case to effect his analysis, then Aristotle would have committed himself to the doctrine. For if it played an indispensable role in either of those discussions, that would suffice for demonstrating Aristotle's commitment. The same holds true for Aristotle's treatment of puzzles deriving from substitution of co-referential terms in opaque contexts, where truth is not guaranteed by co-referentiality alone. Aristotle is himself concerned with opacity in modal and intentional contexts; here HFIO could do significant work for him by explaining what goes awry in such contexts. Hence, his discussion of opacity could provide strong, indirect evidence that he accepts this view.

Aristotle's central discussion of opacity in psychological contexts involves the masked man in *Sophistici Elenchi* 24.[22] In this chapter, Aristotle seeks to diagnose fallacies deriving from the accidental (*para to sumbebêkos*), insisting that one general solution can be given for all such fallacies (*Soph. El.* 179ª26–7). He provides a host of examples of such fallacies, some of which, significantly, are not instances of fallacies involving opaque contexts at all (e.g. *Soph. El.* 179ᵇ4–6). He then moots one type of solution, partially rejects it and partially endorses it, and then fails to supplant it with his preferred alternative. The question we will need to ask in the present context concerns whether his preferred solution makes appeal to HFIO.

How might it? Someone might argue quite correctly:

(1) The man approaching us is wearing a mask.

(2) The man approaching us is the same as Coriscus.

(3) Hence, Coriscus is the same as the man wearing a mask.

---

[22] Matthews (1982, 226–7) rightly remarks: 'There is at least one use to which Aristotle puts the distinction between kooky objects and "straight" ones (that is, substances) that will strike modern readers as centrally important. It is illustrated in Aristotle's treatment of the famous puzzle of the Masked Man.' Still, as I argue in 6. 5, modern readers will be mistaken if they understand this passage as an obvious instance of a problem generated by referentially opaque contexts.

And then she might, appealing to the structure of this argument, offer the following:

(4) The man approaching us is known by Socrates to be wearing a mask.

(5) The man approaching us is the same as Coriscus.

(6) Hence, Coriscus is known by Socrates to be the same as the man wearing a mask.

Yet when we ask whether the man wearing a mask is the same as Coriscus, Socrates responds that he does not know, or even, believing Coriscus to be out of town, that he knows that Coriscus is *not* the man wearing a mask. Hence, something has gone wrong in the move from (4) and (5) to (6).

In diagnosing the error, Aristotle might appeal to HFIO.[23] Because 'the man known by Socrates to be wearing a mask' in (4) designates one object and 'Coriscus' in (5) designates another, non-identical object, there is no grounds for identifying them in (6).[24] There are, indeed, grounds for distinguishing them, if HFIO is true. For HFIO entails that far from being identical, they are distinct, even if they are accidentally the same. If Aristotle reasons this way, then his treatment of fallacies of accident does indeed provide indirect support for HFIO.

From this perspective, it was already tendentious of me to describe the case of the masked man as a case of *opacity*. Opacity is a doctrine about failures to preserve truth when substituting co-referential expressions; and according to HFIO, 'Coriscus' and 'the masked man' are not co-referential. This provides some indication of our first question about a potential appeal to HFIO in the case of the masked man. Aristotle thinks that there is a problem in the inference from (4) and (5) to (6) but that the inference from (1) and (2) to (3) is acceptable. Why should this be so if he adopts HFIO? According to HFIO, the expressions 'Coriscus' and 'the man wearing a mask' are no more co-referential as they occur in (1) and (2) than are 'the man known by Socrates to be wearing a mask' and 'Coriscus' in (4) and (5). It seems, at least initially, that if HFIO blocks the second inference, it also blocks the first. If it blocks the first inference, HFIO blocks too much. Therefore, either Aristotle accepts HFIO and must explain how perfectly acceptable inferences differ from clear fallacies of accident, or at least some of the fallacies of accident discussed in *Sophistici Elenchi* 24 may be cases of genuine opacity.

---

[23] So Matthews 1982, 227–8.
[24] Castañeda (1972) provides a contemporary analogue in 'guise theory'.

It is consequently necessary to look more closely at *Sophistici Elenchi* 24 to determine whether Aristotle ever appeals, explicitly or implicitly, to HFIO. Aristotle opens the chapter by identifying a class of fallacy for which there is to be one general form of solution:

With respect to the accidental (*pros de tous para to sumbebêkos*) one and the same solution <is available> for all cases. For since it is indeterminate when one must predicate <an attribute> to a thing whenever it belongs to its accident, and since in some cases it seems <to people that the attribute does belong> and they say so, and in other cases they deny that it is necessary <that it does so belong>, one must respond similarly to all having drawn <such> conclusions[25] that it is not necessary; and <here> one has to be able to produce an example. All these sorts of arguments depend on accident: (i) 'Do you know what I am about to ask you?' (ii) Do you know the one approaching, or the one who is masked?' (iii) 'Is this your work, or is the dog your father?' (iv) 'Is the number multiplied by a small number a small number?'[26] For it is clear in all these cases that it is not necessary that what is true of the accident is also true of the thing. For all <attributes> seem to belong in only those things undifferentiated with respect to substance and one in substance. (*Soph. El.* 179ª26–39)

This passage raises some difficult questions concerning Aristotle's approach to accidental unity and predication.

First, Aristotle claims that there is some one solution for all fallacies of accident. Yet given the variety of instances he produces, it is hard to see what this could be. Second, cases putatively involving opacity are introduced as special cases of fallacies of accident, not as exhausting the category. At any rate, only the sorts of cases introduced under (i) and (ii) involve opaque *intentional* contexts; and it is initially unclear how they form a class with the cases reported under (iii) and (iv). Finally, Aristotle reserves certain kinds of predications, those which belong to things that are 'undifferentiated with respect to substance and one in substance'

---

[25] Reading *sumbibasthentas* with Poste (1866, 72) on the basis of *D, B*, and *u*, over Ross's (1958) OCT *sumbibasthentos*.

[26] The fallacies Aristotle has in mind here may evidently be expanded to: (i) Since you know what I am going to ask you about, e.g. the nature of goodness, you know about the nature of goodness; (ii) Since you do not know the one who is approaching, or the one who is wearing a mask, and the one who is approaching, or wearing a mask, is Coriscus, you do not know Coriscus; (iii) Since this painting is a work of art, and it is yours, it is your work of art; or, since this dog is a father, and it is yours, this dog is your father; and (iv) Four is a single-digit number; when multiplied by four it is a double-digit number; so four is a single- and a double-digit number. It is difficult to see how the argument recounted in (iv), called by Poste an 'eccentric syllogism' (1866, 156), qualifies as a fallacy of accident. Alternatively, Aristotle's point, as suggested at *Soph. El.* 179ᵇ34–7, may be: a number (e.g. 2), when multiplied by a small number (e.g. 2), yields a small number; so a number, when multiplied by two, yields a small number. This at any rate would make sense of the mistake, about which Aristotle complains, of those who concede the argument but point out that all numbers are both small and large. They have failed to diagnose the problem.

(179^a37–9), as immune from fallacies of accident. Will this reservation be too exclusive, with the result that it treats legitimate inferences as fallacious?[27]

Aristotle's attitude towards these sorts of issues emerges in his diagnosis of the fallacy. He considers first a response by those whose solution proceeds by 'dividing the question' (*diairountes tên erôtêsin*; 179^b7).[28] They think one should say that 'it is possible to know and not to know the same thing, but not in the same respect' (179^b5–7). On their approach, although they know Coriscus but do not know the man coming towards them, they both know and do not know the same object. In one respect, they know Coriscus, for example in so far as he is the man in the dapper garment; in another respect, they do not know him, in so far as he is the man approaching.

Aristotle does not accept this solution and may therefore seem attracted to HFIO. According to this solution, someone knows one and the same entity, Coriscus, but under different descriptions, or under different modes of presentation. Since he rejects this solution, Aristotle must deny that one entity, Coriscus, is known under two descriptions. Hence, he must hold instead that there is not one entity but two, namely, the ones designated by the expressions 'the man in the dapper garment' and 'the man approaching'. Since this is precisely what HFIO entails, Aristotle's rejection of the proposed solution provides confirming evidence that he embraces HFIO.

This argument miscasts Aristotle's position. First, he does not deny that it is possible to know and not to know one and the same thing. On the contrary, he asserts that this may be possible and even allows that it may be a sufficient diagnosis of the problem in some cases of fallacy of accident (*Soph. El.* 179^b13–15, $^b25$–6). His complaint, rather, concerns the generality of the solution. Even if it does explain cases of types (i) and (ii), it will not provide a way of diagnosing cases of types (iii) and (iv) (*Soph. El.* 179^b11–18; cf. 177^b30–4): if a dog is a father and yours, he is not therefore your father, and nothing about how you conceive of him is relevant to appreciating the argument's misstep (cf. *Soph. El.* 178^b14–16). Since they stem from the non-generality of the solution, Aristotle's reservations do not betray any commitment to HFIO. Indeed, in as much as he allows

---

[27] Poste (1866, 158) thinks that 'if admitted [this fallacy] would upset nine-tenths of the syllogisms ever constructed'. This would be true if it held that 'the predicate of a predicate cannot be inferred of the subject unless one of the premises is an essential proposition or even a definition' (158). But the fallacy does not hold this.

[28] The Revised Oxford Translation (Barnes 1984) has, rather mysteriously, 'demolishing the question'. As I argue, Aristotle's point is that the solution they offer is not perfectly general and therefore will not cover every case. Cf. *Soph. El.* 178^b14–16.

that it is possible to know one and the same thing under distinct modes of presentation, Aristotle seems to reject HFIO altogether.

In any case, Aristotle does not articulate a clear and general analysis of the cases he enumerates, and there is some question about whether they are equally instances of some one kind of fallacy.[29] Still, he provides some indication of the nature of the fallacy, and this provides some clue as to why he regards them as instances of the same kind. Consider two cases, one evidently involving opacity, and one clearly not. The first one does not turn on any intentional context:

    (1) This dog is a father.

    (2) This dog is yours.

    (3) Therefore, this dog is your father.

The second, as we have seen, does:

    (4) The man approaching us is known by Socrates to be wearing a mask.

    (5) The man approaching us is the same as Coriscus.

    (6) Hence, Coriscus is known by Socrates to be the same as the man wearing a mask.

What error, if any, do these arguments have in common? Perhaps more important, what error do they commit that arguments of similar structure do not commit?[30]

They may seem to have nothing at all in common. The first argument, as we might think, trades on an equivocal use of 'your', meaning in one context an item you own and the other an item to which you bear some special relation. The second, by contrast, illicitly ignores the opacity of 'knows that'. These would then be distinct types of fallacies, and not appropriately regarded under the same heading. Aristotle sees that some will be inclined to reason this way and thinks they are mistaken. This is why he rejects the solution of those who proceed by 'dividing the question' (*diairountes tên erôtêsin*; *Soph. El.* 179^{b}7), and insists that there should be one solution (*mia lusis*) for all these cases (*Soph. El.* 179^{a}26–7; cf. 177^{b}31, 179^{b}11–13). The problem with this solution is that it is not sufficiently gen-

---

[29] Poste (1866, 156) thinks they are not. He treats some as equivocations, some as fallacies of composition and division, and some as cases of amphiboly. He unsurprisingly concludes: 'We might do well to drop [the fallacy of per accidens] from the list and distribute its contents among the other classes' (158).

[30] The following inference is fine: (1) This Mercedes is a blue car; (2) This Mercedes is yours; hence, (3) This Mercedes is your blue car.

eral, since (i) it will not cover cases which do not involve knowledge ascriptions, and (ii) it will not hold when someone extends the principle 'not to knowing, but to being, or being in a certain condition' (*Soph. El.* 179ᵇ13–15). As Aristotle's example illustrates, it will not suffice to analyse the first argument, since the dog is not in one way your father and in another way not, nor in one way a father and in another way not a father, nor indeed is the dog in one way yours and in another way not yours. The mistake cannot be, he thinks, a simple equivocation at all (*Soph. El.* 179ᵇ38–180ᵃ7).

Rather, Aristotle suggests, in all these arguments equally, 'it is not necessary that what is true of the accident is also true of the thing' (*Soph. El.* 179ᵃ26–7). Aristotle's positive suggestion is initially a bit obscure. His examples help make clear, however, that he means that if two things happen to be true of an object, say that it is a father and that it is your dog, and there is no analytical connection between the two attributes, they are only illicitly combined. Similarly, if someone is known by Socrates to be approaching and to be Coriscus, it does not follow that what is known by Socrates to be approaching is also what is known by him to be Coriscus. This suggests that fallacies of accident occur whenever one incorrectly superimposes one accident of an entity upon another, with the result that the combination yields untoward results. This is, just as Aristotle claims, because 'the combination depends on accident' (*para to sumbebêkos hê sunthesis estin*; *Soph. El.* 180ᵃ4).

Some might conclude even now, then, that HFIO must be driving Aristotle's reasoning. In so far as there is a generic fallacy of accident, it does not turn on opacity as such, even if some examples of what we would call opaque contexts count as fallacies of accident. Rather, the fallacy occurs whenever one combines two accidental things illicitly, e.g. the thing which is my dog and the thing which is a father, or the thing which is known by Socrates to be approaching and the thing which is known by Socrates to be Coriscus. The fallacy is a fallacy because it identifies hyper-finely-individuated objects which are non-identical. This shows, it may be concluded, that the hyper-finely-individuated objects are not identical, and hence that Aristotle does accept, for such objects:

$$(\exists x)(\exists y)((Fx \ \& \sim Gx \text{ at } t_1) \ \& \ (Fy \ \& \ Gy \text{ at } t_2) \ \& \ (y \neq x)).$$

If the fallacy is a fallacy only because those who commit it fail to see this, then Aristotle must after all accept HFIO.

This conclusion is mistaken. When he characterizes the fallacy of accident, Aristotle never insists that the entities combined are non-identical. On the contrary, he suggests that there is no object such that it is a dog and your father, because any such combination would involve a combination depending upon accident. If there is no such object, then there is no object to be constructed out of non-identical hyper-finely-individuated objects. Hence, for this example at least, Aristotle's analysis of the fallacy of accident makes no appeal to HFIO, explicitly or implicitly.

It might be countered that Aristotle nevertheless appeals to hyper-finely-individuated objects in passing. Although there is no object such that it is your dog and your father, there are still the objects, your dog and the dog which is a father, or, indeed, the object approaching and the masked man, where presumably they can at least be combined. If these pairs of objects are non-identical, then Aristotle does in the end embrace HFIO.

There is, however, no reason to suppose that Aristotle ever treats the objects referred to with the singular terms 'your dog' and 'the dog which is a father' as non-identical. Nor is there any reason to suppose that Aristotle ever treats the objects referred to by the singular terms 'Coriscus', 'the masked man', and 'the man approaching' as non-identical. Certainly nothing in his recognition of the fallacy of accident as such requires it: to see that accidental properties of an object can be illicitly conjoined or superimposed is not yet to endorse HFIO, since it is not yet to say that such conjunction or superimposition is the conjunction or superimposition of discrete, non-identical entities. Moreover, Aristotle makes clear that the fallacy results not from the superimposition of *objects* on one another, but rather from illicit *predication* (*Soph. El.* 179^{a}35–7). If someone says that a is F, and that a is G, it will not follow that a is an FG, unless we know antecedently of the non-accidental character of a's being F or a's being G. Hence, no general features of Aristotle's discussion commit him to HFIO.

Aristotle recognizes that there may be connections between something's being F and its being G when he claims that not all combinations of attributes yield unacceptable results. He excepts those cases where attributes 'seem to belong in only those things undifferentiated with respect to substance and one in substance' (*Soph. El.* 179^{a}37–9). His point here is that if we know that either being-F or being-G is a substantial or essential predication of a, then we can infer from the premises that a is F and a is G, that a is an FG (e.g. if this car is a Mercedes and this car is blue, then this car is a blue Mercedes; if a man is known by Coriscus to be approaching and a man is an animal, then an animal is known by Coriscus to be approaching). In licensing these kinds of inferences, Aristotle underscores that his point concerns the combinations of accidental predicates and so undermines the suggestion that his treatment of the fallacy reveals him to be receptive to the kinds of hyper-finely-individuated objects that HFIO postulates.

In *Sophistici Elenchi* 24, Aristotle demands a general diagnosis of a kind of fallacy. This diagnosis should be sufficiently general to account for all instances of the fallacy but should not be so strong as to treat unproblematic inferences as fallacious. According to Aristotle, the fallacy of accident occurs whenever one illicitly combines accidental predicates or mistakenly predicates an accident of an object in order to deduce that in virtue of having that accident, the object in question must also have other accidents. He suggests that although the object may indeed have those additional attributes (Coriscus may be the one approaching and the masked man), this cannot be inferred on the basis of the accident in question. As a special case of the general fallacy of accident, Aristotle considers accidents generated in intentional contexts, where opacity precludes certain derivations. In so doing, he treats opaque contexts not as *sui generis* but as a species of a more general phenomenon. Aristotle does not thereby ignore opacity; nor does he therefore presume HFIO. On the contrary, *Sophistici Elenchi* 24 never appeals to HFIO, and so provides no evidence that Aristotle subscribes to such a doctrine.

Consequently, Aristotle's introduction of the masked man in *Sophistici Elenchi* 24 should be treated in ways analogous to his account of generation in *Physics* i. 7. Both passages appeal to accidental unities; and both identify entities under non-equivalent singular terms. Even so, neither does so in any way which appeals, implicitly or explicitly, to HFIO. There is, therefore, no reason to suppose that Aristotle adopts this counterintuitive principle. Indeed, when we appreciate that Aristotle recognizes various forms of oneness and sameness, we also appreciate that the sense in which accidental unities count as unities at all is in an attenuated and derived sense. We then have no reason to suppose that coincidental unities have any ineliminable role to play in his account of change or theory of substance generally.

6. 6 INTRINSIC AND COINCIDENTAL ONENESS AND SAMENESS

Aristotle opens his discussion of oneness by distinguishing between coincidental (*kata sumbebêkos*) and intrinsic (*kath' hauto*) unities (*Met.* 1015^{b}16–17; cf. 1052^{a}19–1053^{b}8, a mostly parallel passage, and *Phys.* 185^{b}7). The first sort includes some we have already met in the *Topics* and *Sophistici Elenchi*, namely Coriscus and the musical thing; the musical thing and the just thing; and musical Coriscus and just Coriscus. Aristotle recognizes a series of instances of intrinsic oneness, including things one by continuity (*Met.* 1016^{b}36), one in substratum (*Met.* 1016^{a}17), one in genus (*Met.* 1016ª24),[31] and one in definition (*Met.* 1016ª32). All of these ways of oneness, he thinks, are captured by some one core notion.

After enumerating the kinds of coincidental and intrinsic oneness, Aristotle generalizes:

In general, where the thought of something is indivisible—where the thinking is not able to separate the essence in either time, place, or account—these are most of all one, and among these, <especially> however many are substances. For on the whole, whatever does not admit of division, in so far as it does not, is said to be one in this way, for example, if something in so far as it is a man does not admit of division, it is one man, or if something in so far as it is an animal <does not admit of division>, it is one animal, or if something in so far as it is a magnitude <does not admit of division>, it is one magnitude. Hence, of things said to be one, most are said to be one because of something else, because they do or have or suffer or stand in relation to something that is one; others, those said to be one in a primary way (*protôs*), are those whose substance is one, one either by continuity, in form, or in account. For we count as more than one those things which are not continuous or whose species or account is not one. (*Met.* 1016ᵇ1–11)

It is noteworthy that Aristotle readily distinguishes most things (*pleista*) called one from a smaller number of things called one in a primary way.

Those called one in a primary way provide a basis for other things being called one. Aristotle thus provides a core instance of oneness in terms of which other, derived instances are also called. It is consequently possible to construct what I will call a *core-homonymous definition* for oneness; it will subsequently prove possible to define derived instances of oneness in terms of it, in the manner required by our earlier account of core-dependent homonymy. The core-homonymous definition is:

> x is one $=_{\text{chdf}}$ x is indivisible in so far as it is not possible to separate its essence in time, place, or account.

The core notion of oneness is indivisibility. Indivisibility, however, admits of various forms. Accordingly, oneness, even in its core formulation, will admit of various forms.

If this is the core of oneness, then derived applications necessarily make appeal to it, in accordance with our final account of core-dependence, CDH_4. That is, all forms of oneness must stand in one of the four causal

---

[31] Probably Aristotle treats being one in genus and being one in substratum as variations of the same kind of coincidental sameness. Cf. *Met.* 1016ᵇ9 and 1017ª4, where Aristotle recognizes only three types of coincidental sameness.

relations to the core notion in such a way that the core notion is asymmetrically responsible for the existence of the oneness of the derived instances. This is precisely what Aristotle suggests in holding that most cases of oneness count as such in virtue of their standing in relation to something that is one. Indeed, the relations cited (making, having, and suffering; *Met.* 1016ᵇ6–7) are just those we expect to qualify as instances of the four causal scheme.

Coincidental cases of oneness are equally derived. Aristotle makes this especially clear when co-ordinating his accounts of accidental oneness and accidental sameness (*Met.* 1017ᵇ27–1018ᵃ3). Here the white thing and the musical thing are the same (e.g. the white Coriscus and the musical Coriscus), because they are coincidentally the same. Each is also one thing but only because, necessarily, they coincide in a unified substance, a unity which meets the core definition specified (*Met.* 1015ᵇ30–1).

In recognizing these various forms of oneness and sameness, Aristotle does not endorse their mere equivocity. Instead, he thinks they are related to core notions which are narrowly and primarily realized only in entities which are indivisible in the appropriate ways. In recognizing these core-dependent homonyms, he nowhere introduces entities as objects beyond those he customarily recognizes in his theory of substance.[32] On the contrary, he exploits his theory of substance in promoting 'one' and 'same' as core-dependent homonyms.

---

[32] For an opposed view, clearly and forcefully expressed, see Lewis 1982. It may be assumed that some other passages have just this import. For example, in *Metaphysics* vii. 5 Aristotle wonders whether there is an essence for pale man, as opposed to man. He decides that it does not: 'whenever one thing is said of another, there is not what is, in essence (*hoper*), some this (*tode ti*), e.g. pale man is not, in essence, some this, if indeed <being> some this belongs to substances alone' (1030ᵃ3–6; reading *tode ti* with A1p in l. 5). Here there may be a temptation to understand Aristotle as reasoning this way: (i) every substance is some this; (ii) pale man is not some this; (iii) hence, pale man is not a substance; (iv) man is a substance; (v) hence, pale man and man are not identical; (vi) but there is a pale man which is the same as the man; (vii) hence, pale man and man must be in some way numerically the same without being identical. What Aristotle seems to be arguing, instead, is that pale man is not an intrinsic unity, and so lacks an essence, and so does not *as such* qualify as a substance. For the reasons we have already seen, this does not entail that there exists a pale man who is non-identical to the man, something which is a complex object composed of a substance and a non-substantial attribute. Rather, the man, in so far as he is a pale man, is not an intrinsic unity. But the pale man is the man. For a similar chain of inferences, consider a man, his psychiatrist, and his wife: (i) his psychiatrist is an impartial listener; (ii) his wife is not an impartial listener; (iii) hence, his wife is not his psychiatrist. The first two premises might well be true, even though the conclusion does not follow, as long as his wife, who is his psychiatrist, can keep her roles distinct. We should not be compelled to conclude that his wife is not identical to his psychiatrist (she might be surprised to learn that), even if we allow, as we should, that what it is to be a wife is not the same as what it is to be a psychiatrist.

6. 7 THE ONENESS OF THE BODIES

We are now in a position to revisit an issue which puzzled us earlier.[33] Aristotle forestalls deep worries about hylomorphism by distinguishing organic and non-organic bodies. This distinction seemed both *ad hoc* and profligate. It is, however, in no way introduced merely to answer potential worries about hylomorphism: it is motivated by FD and is defensible to the extent that Aristotle's general account of functional determination is defensible. It can now be seen that Aristotle can call the body both one and not one. It is not one, since by Leibniz's Law, non-organic and organic bodies must be distinguished. One is, and the other is not, necessarily alive. Still, these are one, and not merely coincidentally. Rather, they are one in an intrinsic sense, because their 'substratum does not differ in species' (*Met.* 1016ᵃ17). As Aristotle glosses this form of intrinsic oneness, he notes that there is no difference in substratum when the species is indivisible with respect to sense perception (*Met.* 1016ᵃ17–19). In these sorts of cases, we call things one because there is no perceptually discernible difference in their substratum. Even so, as the implied contrast suggests, we may be able to discern a difference in thought. This is precisely the case with organic and non-organic bodies.

This way of proceeding highlights Aristotle's willingness to draw distinctions in philosophical contexts that prove useful, even if they have no independent confirmation in other contexts, including perceptual contexts. When we say that the organic and non-organic bodies are and are not one, we do not contradict ourselves because we implicitly appeal to different forms of oneness, as required by context. Nor do we imply that there is not a core notion of oneness. Core-dependent homonymy here offers unity in multiplicity.

6. 8 CONCLUSIONS

Aristotle sometimes refers to one object in several different ways. In this, he is hardly idiosyncratic. He is, however, keen to combat the inferences some, especially the Sophists, draw from the varieties of reference. He does so by diagnosing the fallacies in sophistical arguments in terms of coincidental oneness and sameness. In offering these diagnoses, however, Aristotle never relies on an ontology of hyper-finely-individuated objects. Instead, he relies, in a wholly defensible manner, on the homonymy of

---

[33] See 5. 7.

oneness and sameness. Once it is appreciated that 'one' and 'same' are core-dependent homonyms, it becomes clear that Aristotle has at his disposal a clear way of recognizing that two things can be one without being identical. He can therefore conclude, defensibly, that organic and non-organic bodies can both be and not be one. If the conclusion that these bodies are discrete is required by Leibniz's Law, nothing problematic will result. More to the point, nothing in this conclusion will require Aristotle to adopt any version of a doctrine of hyper-finely-individuated objects: the bodies can fall under different sortals while being one.

7

The Meaning of Life

7. 1 THE TOOL

The right sort of large firm might advertise an especially useful android for sale on the open market. What makes this android especially useful, the advertisement proclaims, is its amazing *lifelikeness*. According to the firm's publicity agents, the androids in this line 'think, speak, and act like humans—except better, and, needless to say, faster'. Perhaps to assuage concerns about the propriety of buying and owning such a creature, the firm names their product *the Tool* and closes its announcement with the assurance, 'The Tool is essential for commodious living: it performs any domestic task you like with alacrity; indeed, it fulfils every human function but life itself.'

Before buying, we might look beyond the declamations of the advertising hype to determine whether the Tool in fact engages in the full range of characteristically human activities. It thinks. Is the Tool conscious? It speaks. Could the Tool conduct my lectures for me? It acts. Does the Tool consider what is good for itself and devise strategies for reaching its desired ends? It is a mostly silicon-based machine. Presumably the Tool cannot produce (or reproduce?) human offspring? We might, in short, wish to probe the easy assurance that the Tool is not alive.

Some are surprisingly comfortable with the judgement that the Tool is not alive. After all, the Tool is a *mechanism*. Indeed, the Tool is a *programmable* mechanism. It was developed in the laboratory for a definite purpose; its physical states were contrived by conscious designers to enable it to serve that purpose; and it may be turned off at night to conserve its batteries. Of course it might pass a sort of Turing test for life—but that just illustrates the bankruptcy of the Turing test. The case is merely a sophisticated variant on a simpler case, where our intuitions are not so muddy.[1] If after showing you my garden, I reveal that the tulips are in fact powered by small internal machines designed to make them gradually larger and to have them open at the appropriate times, you might

---

[1] The case of the mechanical flower is adapted from Ziff. See n. 2.

well conclude that the tulips, unlike the roses (standard), are not alive. The Tool is like that. It merely gives the appearance of being alive.

This, at any rate, is the sort of conclusion drawn by some contemporary theorists.[2] Indeed, one might argue that we already know the answers to our most important questions about the Tool. We wish to know whether it is conscious. Certainly it cannot be conscious if it is not alive; for, necessarily, x is conscious only if x is alive. Further, we already know that the Tool is not alive. For its internal structure reveals it to be a mechanism no less than the internal structure of the mock tulip reveals the mock tulip to be a mechanism. Hence, the tool is assuredly not alive, since no mechanism, however complex, is alive. If we add further that, necessarily, if x thinks, it is possible for x to be conscious, then we see that the firm's advertisement is positively incoherent. It promotes the Tool as non-living but as thinking. No such creature is conceivable. No matter: our main scruples concerned whether or not the Tool was alive, and if the advertisement overstates the Tool's abilities, then that is the way of advertising. We are, at least, able to purchase the Tool without fear of moral reproach.

7.2 ARISTOTLE AGAINST DIONYSIUS ON LIFE

This string of inferences is bound to seem too quick. However that may be, it is safe to say that it reflects a certain conception of life, which we might call *structuralist*. Structuralists hold that internal structure is definitive of life, in the sense that questions about whether a certain entity is alive are to be settled by examining the entity's structure. Typically, when determining whether something is alive, structuralists will wish to determine whether it has the kind of matter thought of as organic, rather than the sort which we use for producing mechanical devices. By examining whether a thing's internal components are mechanical, structuralists

---

[2] According to Putnam (1975, 402), this is the view of Paul Ziff. 'Ziff wishes to show that it is false that [a robot] Oscar is not conscious. He begins with the undoubted fact that if Oscar is not alive he cannot be conscious. Thus, given the semantical connection between "alive" and "conscious" in English, it is enough to show that Oscar is not *alive*. Now, Ziff argues, when we wish to tell whether or not something is alive, we do *not* go by its *behaviour*. Even if a thing looks like a flower, grows in my garden like a flower, etc., if I find upon taking it apart that it consists of gears and wheels and miniaturized furnaces and vacuum tubes and so on, I say "What a clever mechanism", not "What an unusual plant". It is *structure*, not *behaviour* that determines whether or not something is alive; and it is a violation of the semantical rules of our language to say of anything that is clearly a mechanism that it is "alive".'

profess to be able to determine whether that thing counts as a living organism.

Structuralism is unacceptably limited. In many cases, perhaps in all interesting cases, a distinction between the organic and the mechanistic will be hard to apply or enforce. More important, structuralists are apt to be biologically parochial in their judgements. If we do not know in advance whether some alien entity is alive, and we discover that it is mechanical, at any rate in our terms, because it is driven by internal chips, wheels, and pulleys, we ought not to be constrained to judge that it is not a living creature. Perhaps, for example, its species simply evolved that way in an alien environment. Or perhaps it is a member of a race of beings which, for reasons of health maintenance, gradually replaces the organic parts of its youth with comparatively sterile, easily serviceable mechanical bits. For these sorts of reasons, judgements about life ought to move beyond judgements about structure. Minimally, they ought to include questions about the behaviours and abilities which an entity manifests. Perhaps they should also consider questions about the history and evolutionary pattern of the species to which the entity belongs.

Despite his detailed biological research, or perhaps because of this research, Aristotle shows a healthy disregard for structuralism. This disregard manifests itself in many ways, including notably in his criticisms of Dionysius's univocal view of life:

This occurred also with Dionysius's definition of life, if indeed this is 'movement of a creature which is nourished, when it <viz. this movement> is naturally present within it'. This belongs to animals no more than to plants. But life seems not to be spoken of according to one form, but belongs in one way to animals and in another way to plants. (*Top.* 148ᵃ26–31)

It is worth stressing the character of Aristotle's criticism. He thinks Dionysius erred in being overly general. The problem is not that what Dionysius says is false because it inaccurately characterizes living entities; Aristotle does not object on this score. Rather, he insists, life is not spoken of 'according to one form', but belongs in different ways to plants and animals. Presumably, there are distinct forms of life correlated to the different types of souls Aristotle recognizes (*An.* 413ᵃ30–ᵇ12). If life belongs to various types of living things according to different forms, then there is no one form of life which all and only living things have in common.

So far Aristotle's criticisms reflect a standard appraisal of non-univocity. It is noteworthy that he motivates his appeal to non-univocity by expecting something which belongs in different ways to different kinds of living beings to resist univocal analysis. That is, he claims that life belongs in different ways to different living beings; and he infers from this that life 'is spoken of in many ways'. He does not, however, infer that there are a variety of kinds of life with no connection to one another. In this sense, even while motivating non-univocity, Aristotle evidently presumes association. Aristotle thinks that Dionysius's definition of life is unacceptably simplistic, but he does not argue this on the basis of a radical equivocity of 'life'.

In objecting in this way, Aristotle intends two related criticisms, one concerning the content of Dionysius's proposal, the other its form. Aristotle first of all thinks that Dionysius's definition is inadequate because it does not capture the distinctive forms of life which different sorts of living things realize. In this sense, Dionysius is overgeneral and so provides too thin an account. Consequently, and this is Aristotle's second complaint, Dionysius has the wrong form of definition. Because of its complexity, 'life' requires a non-univocal account; because of its co-ordination, 'life' equally requires an associatively homonymous account. Dionysius misses this by proposing a mono-dimensional, univocal account.

In recounting these distinct types of complaints, I do not yet endorse Aristotle's criticisms of Dionysius. After all, a defender of Dionysius may respond in either of two ways. First, he may simply deny the non-univocity of life. He may, for instance, simply deny that life belongs 'in one way to animals and in another way to plants'. He may fully recognize that animals and plants engage in different sorts of activities as living beings; he will simply deny that their differing activities pertain to what it is for them to be alive. Tigers and snakes engage in different sorts of activities; but they are, according to Aristotle, synonymously animals. Why are they not similarly synonymously alive? And if they are synonymously alive, why will it not be possible for snakes and trees to be synonymously alive? At any rate, their engaging in disparate sorts of activities ought to present no impediment. Moreover, a defender of Dionysius may respond in a second way. She may concede Aristotle's first criticism without granting the second. That is, she may agree that the content of Dionysius's definition is wrong without also agreeing that its general form is incorrect. In responding this way, she will rightly point out that we cannot infer non-univocity from the falsity of a particular univocal proposal. Perhaps Dionysius has simply failed to identify the defensible univocal account.

7.3 ARISTOTLE ON LIFE: A FIRST APPROACH

In order to meet these rejoinders, Aristotle will need to move through the stages of establishing core-dependent homonymy. First, he will need to establish non-univocity; second, he will need to establish association; and, finally, he will need to establish core-dependence without inadvertently relying on a second-order univocity. This last concern will become especially pressing when we examine Aristotle's own positive account of 'life'.

Aristotle introduces questions about the nature of life into the *De Anima* by recognizing their relevance to questions about the nature of the soul:

> Let us say, then, in taking up a new starting point for our inquiry, that what is ensouled is distinguished from what is unensouled by living. But living is spoken of in many ways, and if even one of these belongs to something, we say that it is alive, that is: thought; perception; motion and rest with respect to place; and further motion with respect to nourishment, decay and growth. For this reason all plants too seem to be alive. (*An.* 413ª20–6)

Aristotle's remarks here are in some ways to be expected. They recall his criticisms of Dionysius by re-asserting the multivocity of life. They augment those criticisms by rather boldly introducing a disjunctive set of conditions for life. If anything has at least one of the properties or abilities mentioned, it counts as alive.

In this last respect, Aristotle's remarks may seem surprising. He offers what seems, at least initially, to be a disjunctive definition of life:

x is alive $=_{df}$ (i) x thinks; (ii) x perceives; (iii) x moves (itself) in space; (iv) x takes on nourishment; (v) x decays; or (vi) x grows.

This definition is inadequate in any number of ways. I mention three central problems.

First, some of the disjuncts are not, as stated, sufficient for life. Something can move without being alive; something can even move intrinsically without being alive, so long as no impediments preclude it from doing so (*Phys.* 255ᵇ20–256ª2). I have anticipated this objection by adding 'itself' into the *definiens*. With this addition, we have a sufficient condition. But now, and this is the second issue, we have a problem of circularity. If something is alive only if it moves itself, and only those things which are alive qualify as self-movers, then trivially only self-movers are alive. We have not given an illuminating, non-circular account of 'life'. The same holds true for others among the disjuncts. All natural objects decay. Some non-living natural objects grow in the sense of becoming larger. We do not want decaying cigarette filters to count as alive; nor do we think that balloons expanding in response to changing atmospheric pressure qualify as living beings. If we in turn insist that the sort of growth and decay we have in mind are the sorts suffered only by living beings, we have once again lapsed into uninformative circularity. Aristotle's disjunctive definition, if it is such, is not even as promising as Dionysius's univocal account.

A third problem stems from the form of the definition. Aristotle should not be content with disjunctive definitions of this sort. To begin, disjunctive definitions are as such suspect, because they do not specify whether any other disjuncts may be forthcoming; they do not, that is, carry closure clauses. They are therefore incomplete and thus inadequate. If it is to be presumed that the disjuncts listed exhaust all the possible sufficiency conditions for life, then Aristotle ought to be in a position to provide a reason why this is so. He must be able to say in a general way what all of these disjuncts have in common such that nothing else could qualify as a sufficient condition for life. Were he in a position to do so, Aristotle would not need to rely on this form of definition.[3]

Fortunately, Aristotle need not be regarded as introducing a disjunctive definition of life. Indeed, we can hardly expect Aristotle to think of himself as *defining* life in this passage. This list of disjuncts does not constitute a canonical Aristotelian definition in terms of genus, species, and differentia.[4] Moreover, if, as Aristotle's criticisms of Dionysius suggest, life is to be not merely a homonym but a core-dependent homonym, nothing about this set of disjuncts reflects this fact.

All of this suggests that Aristotle does not here offer an account or definition of life. Rather, as he says, he merely introduces *ascription conditions* for our regarding something as alive. Rather than offering a definition, Aristotle merely specifies conditions under which we are immediately justified in regarding something as alive. Why he should do so will become clear after we examine his account of life as a core-dependent homonym.

7.4 ARISTOTLE ON LIFE: A SECOND APPROACH

A reading of the *De Anima* shows that Aristotle does not rest content with ascription conditions for life. Instead, he goes further to specify what it is that all of the disjuncts have in common such that they are criteriological of life. As he continues to explore life in the *De Anima*, Aristotle adds, more fully:

[3] Perhaps a parallel from philosophy of mind is useful here. In response to charges that type-type identity theories were hopelessly parochial, some theorists moved towards disjunctive definitions (e.g. pain $=_{df}$ (i) c-fibre firing, or (ii) silicon state c, or (iii) alien state z, etc.). These definitions were rightly regarded as unsatisfactory, largely for the reasons mentioned in the text.

[4] See *Top.* i. 5, esp. $101^{b}37–102^{a}16$.

Consequently, we must speak first of nourishment and generation. For the nutritive soul belongs to <living things> along with all the other <faculties of soul>,
and it is the first and most common capacity of the soul, in virtue of which life
belongs to all living things ... The most natural function for a living thing, at least
for those which are complete and not mutants or which have a spontaneous <form
of> generation, is to make another like itself, an animal making an animal and a
plant a plant, so that they may partake of the eternal and divine in so far as they
are able. For everything desires this, and everything does whatever it does naturally for the sake of this. (*An.* 415ª22–ᵇ2)

In this passage, Aristotle advances an account of life which offers primacy
to one life activity, namely reproduction. This activity is central to life and
structures other activities, all of which are done 'for the sake of this'
(*ekeinou heneka*; *An.* 415ᵇ1).

This primacy has suggested to some that for Aristotle life is essentially
what I will call *biological life*, that is, life in a species with the primary function of reproduction.[5] This seems reasonable, in so far as Aristotle suggests that both the capacity and the impulse for reproduction belong to *all*
living things and that the activities of the most common form of soul, the
nutritive soul, subserve this end (*An.* 415ª25; cf. *GA* 730ᵇ34, *Pol.* 1252ª26).
So perhaps Aristotle supposes that having a nutritive soul *is* being alive
and that being alive is primarily *for* the perpetuation of the species. By
stressing the fundamentality of species membership, and by regarding life
as primarily an impulse for the perpetuation of self and species, some have
attributed to Aristotle a biological definition of life:[6]

> x is alive $=_{df}$ there is a species s, and a psychic power p, such that x
> belongs to s, p is a psychic power for species s, and x can exercise p.

This biological account of life highlights Aristotle's deep commitment to
the teleology of life by making it definitional of life that something exercise its psychic powers for the purpose of perpetuating its species.

This definition is an improvement over the first, because it avoids its disjunctive character. Even so, it remains problematic in some of the same
ways. Most notably, the problem of circularity will re-assert itself, since
something manifests a *psychic* power only if it is ensouled, that is, only if
it lives.[7] The problem is not merely that 'being alive' and 'having psychic

---

[5] I have in mind especially Matthews 1992. Although I am critical of Matthews's final
account, I am nevertheless indebted to his engaging discussion.

[6] This is Matthews's final account (1992, 191). As he says, 'Still, "is alive" does not *mean*
(on this reconstruction of Aristotle) "is capable of self-nutrition". What it does mean is "an
exercise of a power such that members of the organism's species may survive".'

[7] Matthews (1992, 192) responds to a related circularity problem, concluding that it is
not vicious. I have not found his responses convincing. In any case, the circularity problem
I mention in the text is distinct from the one Matthews entertains.

powers' will be co-extensive for Aristotle. Any account of Aristotle on life must recognize this. Rather, an account of life in terms of psychic powers implicitly appeals to the notion of being alive. For the having of a psychic power is simply the manifestation of a capacity of living things as such.

There may be some question as to whether one ought to regard this circularity as vicious. However that issue is resolved, there are two additional problems for this definition. The first is that by making life species-dependent, this definition is, so to speak, too biocentric. According to the biological conception, life is to be understood in terms of an entity's belonging to a species and manifesting the psychic capacities characteristic of that species.[8] This cannot be Aristotle's account of life, since it precludes the possibility of a life for god, and Aristotle's god is surely alive:

If, then, god is always in the condition <i.e. contemplation> that we sometimes are,[9] then this is marvellous. And if god is in a better condition, then this is still more marvellous; but god is in this condition. And life, to be sure, belongs <to god>; for life is the actuality of mind, and this actuality is god. And god's supreme and everlasting life is intrinsically actuality. Hence,[10] we say that god is a living being, everlasting and supreme, so that life and a continuous and everlasting lifetime belong to god. For this is god. (*Met.* 1072ᵇ24–30)

Aristotle emphasizes several times over that god is alive, that god's life is everlasting and good; and he rightly relies on a commitment to god's thinking to support this attribution. God's thinking is sufficient for his being alive.

God thinks and so is alive. Yet god is not a member of a species; nor does god think in order to reproduce or perpetuate any species. Hence, the biological account of life misses something Aristotle regards as central to life, namely, the striking locution that 'life is the actuality of mind'. Accordingly, the biological account is inadequate as an account of Aristotle's conception of life.

There is a second problem. It is unclear why the biological conception qualifies as homonymous. Recalling the second rejoinder of Dionysius's defender, that it is possible that Dionysius merely advanced the wrong univocal account, one might have offered the biological definition as the right univocal account. This latter account holds that life is the manifestation of a species-indexed psychological power. While it is true that different species have distinct psychic powers, this does not suffice for non-

---

[8] I here pass over the question about mutants and freaks which are not members of species, as well as the question of naturally sterile animals. Aristotle seeks to exclude them from consideration at *An.* 415ᵃ27.

[9] I omit *eu* with the first hand of J.

[10] I accept *dê* from Themistius, with Ross, over the mss. *de*.

univocity. Recall that second-order definitions, like causal accounts of poison, nevertheless count as univocal. (Something's being second order does not suffice for non-univocity, since one will insert the same account of *F*-ness, e.g. a substance capable of destroying or harming the life of an animal or plant, in each occurrence of '*a* is *F*' and '*b* is *F*', e.g. 'Hemlock is a poison' and 'Arsenic is a poison'.) If this is correct, then the biological account misses something crucial to Aristotle's criticism of Dionysius, namely that he erred by not merely endorsing the wrong univocal account, but by presuming univocity in the first place.

7. 5 THE NON-UNIVOCITY OF LIFE

Although incompatible with the biological conception of life, Aristotle's conception of god's life coheres with the discussion of the *De Anima* in two important respects. First, Aristotle makes clear there that he means to restrict his discussion to the sort of life mortals enjoy.

It is clear, then, that there would be one account of both soul and figure in the same way. For in the one case there is no figure beyond triangle and those that follow in order <quadrilateral, etc.>; in the other case, there is no soul beyond those mentioned. There might be a common account for figures, which fits them all; but it will not be peculiar to any one of them. Similarly, <there might be a common account> for the souls mentioned. For this reason, it is ludicrous to seek a common account for these and other cases, which will not be an account peculiar to any of the things which exist, nor in accordance with the proper and indivisible species, while omitting this sort of thing. (Things falling under soul are similar to what occurs concerning figures; for what is earlier in the series for both figures and ensouled belongs in potentiality to what follows, e.g. the triangle in the square and the nutritive capacity in the perceptual.) Consequently, one must inquire into each individually, what the soul of that is, for example, what the soul of a plant or a man or a beast is. One must look at the cause in virtue of which <things> stand in this serial way. For the sensory capacity does not occur without the nutritive capacity; but the nutritive capacity is separated from the sensory capacity in plants. Again, none of the other senses belongs <to living things>; but touch belongs without the others. For many of the animals have neither sight nor hearing nor the sense of smell. And among those things capable of sense, some have a capacity for local motion and some do not. Finally, and fewest in number, are those which have reckoning and thought; for among mortal beings, those to whom reckoning belongs, so too do all the rest <of the psychic capacities>, whereas reckoning does not belong to each of the others. Rather, imagination does not belong to some, while others live by this alone. There is a separate account about the contemplative mind. (*An.* 414^{b}20–415^{a}12)

Souls are hierarchically structured, but only among mortals. So, in any mortal being, we can infer the presence of a sensory capacity from the presence of mind; and from the presence of sensory capacity, we can infer the presence of the nutritive capacity. This, however, does not hold universally true, as Aristotle indicates. It is possible that there be a mind, a living being, which lacks all the lower orders of soul. It would therefore be a mistake to ascribe to Aristotle a view which precluded this possibility—a possibility, one may add, which Aristotle regards as actual.

Second, the postulation of a living though non-biological being is continuous with the commitment to the non-univocity of life in the *De Anima* (413ᵃ22); and to a certain extent, it helps support that contention. For in some living beings, higher-order capacities require lower-order capacities; in other living beings, this is not so. Some living beings exercise their capacities primarily for the purpose of self-preservation and species propagation. Other living beings have no need of self-propagation: they are everlasting. Hence, the conditions of life are deeply variable across living beings.

Even so, one might hold a brief for Dionysius by pressing the univocity of life in principle.[11] Conditions of life may vary. But this does not entail that life itself varies. Indeed, Aristotle seems to allow that there could be a common account for souls, just as there is for figures (*An.* 414ᵇ22–4), his main scruples are against the search for a common account in the absence of an account peculiar to the various kinds of figures and souls (414ᵇ26–7).[12] So again, while the appearance of non-biological living beings tells against a particular univocal account, it does not by itself establish non-univocity.

As we have already seen,[13] neither the substitution test nor the test of opposition of the *Topics* i. 15 provides any immediate help to Aristotle on this score. Thus, if we provide paraphrases of 'is alive' in:

(1) Socrates is alive.

(2) Pavlov, the dog, is alive.

(3) My florabunda rose bush is alive.

we might well end up without any indication of non-univocity, e.g. '*x* engages in end-directed psychic activity for the purpose of self preservation and species propagation'.

---

[11] For a sophisticated attempt to hold out for a kind of univocity as against some express doubts as to whether that is possible (given in n. 18 below) see Bedau 1996, which offers an account in terms 'supple adaptation'.

[12] Here I agree with Hicks 1907, 336–7, note to 414ᵇ27.

[13] See 2. 4.

Pressing the point, we might add:

(4) God is alive.

A modified defender of Dionysius might still insist that a univocal para-phrase can be given, perhaps '*x* engages in end-directed psychic activity for the purpose of self-realization'.

As for the test of opposition, presumably '*x* is dead' will suffice equally for (1)–(4).

Faced with this defence of Dionysius, Aristotle need not simply acqui-esce. Evidently 'life' is a case where homonymy 'trails into the accounts themselves unnoticed', where 'for this reason one needs to look into the accounts' to uncover the unnoticed non-univocity (*Top.* 107^{b}6–8; cf. 134^{a}27). Given that there is deep and shallow signification, it is possible that shallow paraphrase will fail to turn up what deep paraphrase, that is, paraphrase supported by deep signification, will reveal.[14]

Aristotle provides a reason for believing that 'is alive' signifies different properties in these applications, even if this is not obvious on the basis of respectable, but shallow, substitution tests. His argument emerges in the *De Anima*, in a passage introducing the multivocity of cause (*aitia*) and source (*archê*) for the purpose of determining how the soul is the cause and source of the living body.

> The soul is the cause and source of the living body. But these <cause and source> are spoken of in many ways. Similarly, soul is a cause according to the ways delin-eated, which are three: it is a cause <as> the source of motion, <as> that for the sake of which, and as the substance of ensouled bodies. That it is a cause as sub-stance is clear, for substance is the cause of being for all things, and for living things being is life, and the soul is also the cause and source of life. (*An.* 415^{b}8–14; cf. *PN* 467^{b}12–25, *Phys.* 255^{a}6–10)

Aristotle thinks that 'for living things being is life' (*to de zên tois zôsi to einai estin*); and when he asserts this he provides grounds for his commit-ment to the non-univocity of 'life'.

Aristotle sometimes says that *F*s and *G*s are the same, even though being for an *F* is not the same as being for a *G* (*An. Pr.* 67^{b}12, *Top.* 133^{b}33, *An.* 431^{a}9). (His characteristic locution is *to einai F*$_{\text{dative}}$ is not the same as

---

[14] This is the issue that divides Wiggins 1971 and Alston 1971 on the question of how to determine whether a word has more than one meaning. Wiggins points out that in some cases paraphrase will fail to turn up ambiguity. (His argument is unconvincing, but his con-clusion is correct.) Alston counters that paraphrase seems to work in a wide variety of cases. From my perspective, shallow paraphrase suffices for shallow meaning, extending to most contexts of linguistic competence; but only deep paraphrase suffices for philosophically dis-puted contexts. Deep paraphrase is delivered only by philosophical analysis.

to einai G_{dative}, that is, being for an *F* is not the same as being for a *G*). Usually in such contexts, he means that though '*F*' and '*G*' are coextensive, what it is to be an *F* is not the same thing as what it is to be a *G*. For example, fully actual objects of perception are coextensive with actual episodes of perception; but what it is to be an object of perception is not the same as what it is to be an episode of perception. In speaking this way, Aristotle means to highlight what is essential to *F*s qua being *F*.[15] In identifying life and being for living things, he therefore means to point out that living things are essentially alive and that their essence is their living.

This should be unsurprising. We have already explored the implications of this way of thinking for the homonymy of the body.[16] Now, however, another facet of Aristotle's thinking about the soul and life can be emphasized. If Socrates' essence is to be alive, and god's essence is to be alive, and Pavlov's essence is to be alive, and if their essences are distinct, then life is also distinct for them. The identification of essence and life for living things grounds Aristotle's commitment to the non-univocity of life.

His argument for the non-univocity of life, then, is this:

(1) For living things, to be is to be alive.

(2) Being for *F*s is what is essential to *F*s.

(3) So, in the case of living things, essence is identical with life.

(4) The essence of Socrates is not identical with the essence of Pavlov, the dog; nor is it identical with the essence of my rose bush; nor again is it identical with the essence of god, which is complete actuality.

(5) Hence, life is not the same for Socrates, Pavlov, my rose bush, and god.

(6) Hence, 'life' is non-univocal.

If we accept (1)–(3), then (4) becomes the crucial premise. Now, however, Aristotle can rely on his analysis of the essences of each of the kinds of entities mentioned. If these are distinct, then (5), and so (6), will follow directly.

If we follow Aristotle's standard definitions of the various entities mentioned, we will be able to reapply the tests of *Topics* i. 15, but with different results. For now, relying on deeper paraphrases, we have:

---

[15] Aristotle sometimes uses a corresponding expression *to ti ên einai*, on which see Owens 1951, 180–8; Ross 1924, i. 127; and Trendelenburg 1828, 480.

[16] See Chapter 5.

(1') Socrates engages in activities that are rationally structured.

(2') Pavlov, the dog, engages in perceptual activities that are not rationally structured.

(3') My florabunda rose bush engages in non-perceptual nutritive activities.

(4') God engages in purely actual contemplation.

The test of opposition would also yield the relevant differences. For Socrates to be dead is for him not to engage in rationally structured activities. If Socrates were to become a vegetable, Socrates would cease to exist.[17] Aristotle's views about life imply essentialist views Aristotle elsewhere defends; his views about life are defensible if his essentialism is defensible.

7.6 THE CORE OF LIFE

Aristotle thinks of life as non-univocal because 'life' signifies distinct properties in distinct applications. If so, 'life' cannot be synonymous across its various applications. At the same time, life is a clear candidate for association: one is bound to believe, even if one accepts its non-univocity, that 'life' is not a mere or accidental homonym. This is why defences of Dionysius, if not correct, are reasonable and well motivated.

Though he does not say so explicitly, Aristotle suggests that life is a core-dependent homonym. It is therefore incumbent upon him to specify its core and to explain how non-core instances stand in appropriate relations to that core.[18]

Aristotle meets this obligation in the *Metaphysics*. When defending the claim that god is alive, Aristotle asserts, straightforwardly, that 'life is the actuality of mind, and this actuality is god' (*hê gar nou energeia zôê, ekeinos de hê energeia*; 1072b26–7). Two points are noteworthy. First, Aristotle's identification here is direct and striking: life is the actuality of

[17] This is also evidently the force of *Metaphysics* 1006a14–15.

[18] Some researchers in the field of artificial life have implicitly doubted whether core-dependent homonymy is possible once non-univocity is granted. A typical reason is offered by Boden (1996, 1): 'There is no universally agreed definition of life. The concept covers a cluster of properties, most of which are themselves philosophically problematic: self-organization, emergence, autonomy, growth, development, reproduction, evolution, adaptation, responsiveness, and metabolism . . . It is not even obvious that . . . life is a natural kind. In other words, "life" may not be a scientifically grounded category (such as water, or tiger), whose real properties unify and underlie the similarities observed in all those things we call "alive".'

mind. Second, Aristotle cannot possibly hold that life for all living creatures is the actuality of mind. As he insists repeatedly, not everything
which is alive has a mind. Indeed, not everything which is alive has perception. It is, therefore, not possible that Aristotle means to restrict life to
minded creatures.

Instead, because he thinks that life is non-univocal but that the various
applications of life are focally related, Aristotle asserts this identification
as the core of life. If we understand that the actuality of mind in its purest
form is uninterrupted contemplation, that is, the actuality of mind is
the actualization of the highest and best objects of thought (*Met.*
1072^{b}17–19), then we see that the core of life is a form of enriched intentional activity. Aristotle suggests as a core-homonymous definition of life:

$$x \text{ is alive } =_{\text{chdf}} x \text{ is an intentional system.}$$

Aristotle's god is supremely and completely an intentional system.
Consequently, this god is not only alive, but alive in a way which makes it
the 'one source' (*mia archê*; *Met.* 1003^{b}6) of all life.

The notion of an intentional system upon which Aristotle relies is itself
open-textured. An intentional system is a system whose nature is accurately explained and whose behaviour is reliably predicted by understanding it as engaging in certain forms of end-directed behaviour. It is also a
system whose activities can be assessed in part by the degree to which it
attains its goals, and indeed a system whose goals provide a normative
standard of its being a good instance of its kind. We should not, however,
infer univocity on this basis. For the variety of goals and the variety of
systems employed for attaining those goals differ across the spectrum of
living things and so mark out different forms of life.

Each of these points requires elaboration. God is a teleological system
whose goal is serene and seamless contemplation. A plant lacks this goal.
Both nevertheless qualify as intentional systems, because both have goals.
Moreover, the intentional systems of different forms of life are distinct.
Humans have fully representational mental systems involving imagination
(*An.* 427^{b}27–428^{b}10), indeed imagination of a rational sort (*An.* 434^{a}7).
This sort of capacity involves the storing and manipulation of images and
other forms of representation.[19] Animals have this capacity in a limited
way, and plants lack it altogether. For all these systems, though, the end
provides a standard of assessment for its successes and failures.
Something counts as a good of its kind to the degree that it realizes its end,
where its ends are not available for deliberation. Rather, a living thing's

---

[19] On imagination, see Wedin 1988 and Schofield 1975.

ends partially determine what kind of thing the entity in question is, in line with FD, the functional determination thesis.[20]

If 'life' is a core-dependent homonym, all living things must stand in one of the four causal relations to the form of intentional system god is. Surely god is not the material cause of other living things; nor is god in Aristotle's metaphysics an efficient cause of the life of all living things. Rather, given that all living things are kinds of intentional systems, it follows that they stand in formal causal relations to the core notion.[21] Now, however, we can appreciate the force of Aristotle's allowing formal causation without univocity.[22] For though god's life is a formal cause of the lives of other creatures, the forms thus realized are already distinct. The result will be that living things bear formal causal relations to a pure and complete enriched intentional system without themselves being pure, complete, or enriched.[23] Non-living things may resemble such systems in various ways, but they will not have, in Aristotle's terms, sources of motion within themselves (*EE* 1222^{b}15–30).

This last point is worth exploring in greater detail, since it will be reasonable to object that Aristotle's account is overly generous. We think that heat-seeking cruise missiles, computer-guided navigational systems, and self-monitoring automata of various sorts all count as intentional systems. They, like the Tool, will certainly count as alive, then. If this is a consequence of Aristotle's account, surely his view must be rejected.

Aristotle has two related responses available to him; the first emphasizes the method of homonymous definition, and the second distinguishes between native and non-native intentional systems. In the first case, Aristotle may reasonably respond that two *F*s may be quite similar along some dimension without its being the case that the one's being *F* is a formal cause of the other's being *F*. After all, a statue and a man can be quite similar in appearance, and can both be called 'man', without its being the case that either is the formal cause of the other's being a man. This question is not determined by similarity or resemblance, both in part psychologistic notions, but in terms of whether the accounts of *F*-ness are in each

---

[20] See Whiting 1988 for a defence of Aristotle's reliance on function in determining real kinds. We should not, however, expect functional determination to be an empirical matter in the way in which Whiting suggests.

[21] It may be suggested that god is a final cause on the basis of *Metaphysics* 1072^{b}3. There god is regarded as a final cause of motion. Here, however, we are concerned with the way in which god can be said to be the cause of 'life' for living things.

[22] See 4. 4 on formal causation without univocity.

[23] Human beings engage in full-blown intentional activities, though not completely or everlastingly. This is why Aristotle sometimes points to human intellectual activity as divine (*Met.* 1072^{b}24–8, *EN* 1177^{a}14–19, ^{b}33; cf. *EN* 1153^{b}32, 1174^{a}4, *An.* 415^{a}29).

case the same. If they are, there is synonymy; if they are not, there is non-univocity. If they are homonymously *F*, resemblance will not by itself guarantee that they are associated.

The worry about clearly non-living intentional systems is that they bear more than a resemblance relation to living intentional systems. They equally have ends. And they employ representations in an effort to attain their ends. The missile seeks the object determined by its coordinates and even engages in avoidance behaviour in order to reach its target. Here Aristotle may respond, and this is the second point, that the end in the case of artefacts is clearly derived, whereas the end of a living system is not determined by the design of some conscious agent. Surely Aristotle recognizes that certain sorts of ends, native ends, are not the result of conscious design:

It would be absurd not to think that something has come about for the sake of something, <merely because> one should not happen to see the moving cause making a deliberation. Indeed, even craft does not deliberate. If the ship-building art were in the wood, it would produce in manner similar to nature. Hence if that for the sake of which is in craft, so too is it present in nature. This is most clear whenever a doctor doctors himself; for nature is similar to this. (*Phys.* 199ᵇ26–32; cf. *EN* 1112ᵃ34)

Aristotle's immediate goal in this passage is to point out that the capacity to deliberate is not required for something's qualifying as a system with an end. His argument proceeds by recalling that natural entities move forward toward their ends in just the way craftsmen move forward towards their ends, qua craftsmen. In so speaking, Aristotle accepts a commitment to native, end-directed systems whose ends are not derived from without.

A native intentional system, then, is one whose ends are not derived from the designing activities of external agents. Some may doubt the possibility of such systems.[24] Aristotle is not among them. By accepting such a distinction, Aristotle can plausibly distinguish between native and derived intentional systems; and he can consequently deny life to non-native intentional systems. His distinction will not be in any way ad hoc. Instead, it will be grounded in a broader teleology which understands the ends of native intentional systems in terms of what is good for those systems. It will then be a hallmark of non-native systems that questions

---

[24] Others are comfortable with Aristotle's commitment. In a clear discussion, Dretske (1995, 6–8) accepts what he calls a form of natural representation, which he contrasts with conventional representation: 'When a thing's informational functions are derived from the intentions and purposes of its designers, builders, and uses . . ., I call the resulting representations *conventional*. Representational functions that are not conventional are *natural*' (7). See also Kitcher 1993 and especially Neander 1991.

regarding the good of those systems will have little or no point. It is not good for my hammer to drive nails; and it is not good for the cruise missile to destroy its target. By contrast, it is good for Socrates to contemplate.

If this is correct, non-native intentional systems will resemble native intentional systems without being genuine, intrinsic intentional systems. Aristotle may therefore justifiably treat intentionality as the core of living without also regarding cruise missiles and the like as alive. That said, it will nevertheless be true that it will be at times difficult to determine whether a given system counts as a native intentional system. Indeed, even when all the empirical facts are known, it may be difficult to determine whether a system stands in one of the relations prescribed by FCCP to the intentionality evinced by Aristotle's god. This is, however, not an objection to Aristotle's approach. On the contrary, in so far as it permits the concept of life to be scaled and plastic, Aristotle's commitment to life's core-dependent homonymy is still more defensible. For living beings present themselves in bewildering variety, ranging from complex conscious agents capable of indulging in autonomous contemplation to the feeble simplicity of pond scum. This is a datum to be explained, not a problem for the nature of life.

7.7 CONCLUSIONS

On behalf of Aristotle's account, at least the following can be said. First, it eschews the biocentrism of the biological account. It escapes parochialism by recognizing and embracing the possibility of disparate forms of life. It is, consequently, appropriately comprehensive. It recognizes as alive all plants, all animals, all rational creatures, and, importantly, all creatures, including god, which are native intentional systems or which stand in one of the four causal relations to intentional systems in accordance with CDH_4. Second, this account does not collapse into a mere disjunctive analysis. At the same time, it explains why Aristotle lists the particular disjuncts he does. Aristotle holds that if x thinks, perceives, moves (itself) in space, takes on nourishment, decays, or grows, then x is alive. This is because each one of these activities is sufficient for x's being some form of intentional system. Finally, this account does not lapse into any circularity. One can specify what it is to be an intentional system without appealing to life itself. For an intentional system is a system whose nature is accurately explained by its functions and whose behaviour is reliably predicted by understanding its behaviour in terms of its directing its energies towards those ends. None of these notions appeals to life itself.

Aristotle's account of 'life' as a core-dependent homonym is largely defensible. It is open-textured yet determinate. It structures a series of features often regarded as fundamental to life in such a way as to explain their unity and interconnections. Living things can respond to their environment in self-perpetuating ways, sometimes by representing it, sometimes merely by engaging in immediate, shielding behaviour. Living beings are indeed beings whose behaviour is reliably predicted by ascribing intentional states to them, whether or not these are representational. Living beings are also systems which engage in end-directed, self-moving, purposive behaviour, tending, as Aristotle so often stresses, to engage in autonomous morphogenesis. Whereas the divine life is wholly actual, residing in a state of seemless actualization, a mortal life is a pattern in space and time, one which registers and stores information about its past and future; such a living thing converts energy from the environment into something which serves its self-generating purposes; and this living system is, consequently, stable and forward-moving in the face of minor environmental perturbation. Each of these traits gets life partially right. None of them specifies the core of life. Altogether, they do not capture all living things equally and comprehensively. This is because 'life' is a core-dependent homonym.

Aristotle's account of life reflects these facts about living things without making them definitional of life. This is desirable, since definitions of life should be neither univocal nor disjunctive. Still, we expect our definition of life to register these non-accidental characteristics of living beings. Aristotle's account does so. More important, though, Aristotle's account explains why a single fact resonates through every attempt to define life: the fact that it is possible to ask of every living creature whether it is flourishing.

8

Goodness

8.1 A PROBLEM ABOUT THE NON-UNIVOCITY OF GOODNESS

Aristotle's appeals to homonymy have thus far met with some success. He is justified in treating 'body', 'oneness', and 'life' as homonymous. His treatment of life is an especially fruitful application of homonymy: the framework for core-dependent homonymy which Aristotle develops seems tailored to the richly textured concept of life. It displays both its multiplicity and its order; to some extent, Aristotle's success in this case provides motivation for taking the entire apparatus of core-dependent homonymy seriously.

Given Aristotle's success in establishing core-dependent homonymy thus far, it is reasonable to ask whether his most celebrated and difficult applications, to 'being' and 'goodness', also succeed. In this chapter, I consider the homonymy of 'goodness'. I argue that Aristotle's appeal to the homonymy of goodness is at best a mixed success. Some arguments upon which he relies fail. Others, less often noted, succeed, but only in so far as they establish its non-univocity. Because Aristotle's unsuccessful arguments are required to establish association, his account of the core-dependence of goodness is incomplete. It will be appreciated, however, that this is something which Aristotle himself allows (*EN* 1096^{b}26–31).

I proceed by first placing the homonymy of goodness into its appropriate anti-Platonic context. I next point to some problems for Aristotle's arguments. Using the framework for core-dependent homonymy developed in Part I, I then determine a central adequacy constraint for Aristotle's appeals to homonymy in this area. I consider several attempts to articulate and defend Aristotle's main argument for the homonymy of goodness in *Nicomachean Ethics* i. 6, what I will call the *categorial argument*. I argue that none of these interpretations succeeds, because none establishes non-univocity. I then illustrate a successful argument for the homonymy of goodness drawn from a later Aristotelian philosopher, Thomas Aquinas. This argument is successful in so far as it neatly meets the adequacy constraints laid down for establishing the non-univocity of goodness. In this Aquinas's argument contrasts sharply with Aristotle's, since Aristotle offers no clearly articulated defence of the non-univocity of goodness. Unfortunately, Aquinas's argument appeals to some highly idiosyncratic premises (whose independent merit I do not assess), alien to Aristotle and unavailable to him. I introduce this argument for the sole purpose of determining whether Aristotle has analogous premises or argumentative strategies at his disposal. I argue that he does but that these strategies generate a more limited form of the homonymy of goodness than has traditionally been found in Aristotle.

Taking all of this together, the argument of this chapter exhibits the following structure. Aristotle offers a categorial argument to establish the homonymy of goodness. This argument fails. Its failure is especially problematic, since, if successful, the categorial argument would provide a natural way of articulating the association of the non-core instances of goodness around a core. That is, if the categorial argument could show that goodness is spoken of differently in all the categories but that it is nevertheless spoken of foundationally in one category in the way that substance is supposed to be primary relative to the other categories of being, then goodness would have a natural core application from which non-core instances could be derived in terms of FCCP. If my assessment of the categorial argument is correct, then Aristotle will not have any clear mechanism for establishing core-dependence. Nor even will he have established non-univocity.

Even so, I equally stress that Aristotle deploys a second argument for non-univocity. This argument succeeds. Unfortunately, this argument does not carry the seeds of any clear strategy for establishing association; nor is it clear how the suggestions of the categorial argument might be exported to the conclusions of this argument in such a way that association can be established. Consequently, although he provides a defensible reason for believing that goodness is non-univocal, Aristotle advances no compelling argument for the thesis that it is a core-dependent homonym. This conclusion is, of course, compatible with the core-dependence of goodness, and it even provides the first step in establishing goodness as a candidate for core-dependence. It leaves the job of establishing core-dependence undone, in just the way Aristotle does in *Nicomachean Ethics* i. 6, when he asks, after advancing both the categorial argument and the second, less often cited argument for non-univocity: 'How then is <the good> spoken of? For at any rate it does not seem to be among those things which are homonymous from chance. But is it the case then that all <good things are good> by being derived from one good or by contributing to one good? Or are all good things called good by analogy?' (*EN* 1096ᵇ25–8). The uncertainty expressed here suggests that

Aristotle takes quite seriously the Platonic rejoinder to the categorial argument.

8. 2 ARISTOTLE'S GENERAL ARGUMENT AGAINST PLATO'S ACCOUNT OF THE GOOD

Aristotle's initial argument for the homonymy of goodness is critical. This is reasonable, since he wishes to establish its non-univocity as against Plato's assumption that, in Aristotle's terms,[1] goodness 'is something universal, common <to all good things>, and single' (*koinon ti katholou kai hen*; *EN* 1096ᵃ28). Aristotle's target is, broadly, the Platonistic view that goodness is simple, in that it lacks proper parts; indefinable, because definition requires specifying the parts of a thing in their proper relation to one another; and non-natural, because it is neither identical with entities existing in space and time (or sets of such entities) nor partially constitutive of entities existing in space and time.

Aristotle himself describes the view he assails with greater precision:

(1) The Form of the Good (FOG) is some one universal, common to all good things (*EN* 1096ᵃ23–9).

(2) FOG is separate from good things (*EN* 1096ᵃ34–ᵇ3), where x is separate from y iff x can exist without y.[2]

(3) FOG is eternal or sempiternal (*EN* 1096ᵇ3–5).

(4) Since there is a single science corresponding to each form, there is a science of goodness (*EN* 1096ᵃ29–34).

---

[1] In this chapter I do not pursue the important question of whether Aristotle accurately characterizes Plato's account of the good, as it comes into view in the dialogues or as it may be conjectured to be in Plato's unwritten doctrines. Of course, the question of whether Aristotle's criticisms of Plato on goodness succeed must undertake to answer this question. Here I consider the independent question of whether Aristotle's criticisms against the view he articulates and ascribes to his 'friends'—plausibly regarded as Platonists other than Speusippus—are compelling (*EN* 1096ᵃ13, 1096ᵇ6–7). For the present discussion, I presume that the view described in the text is meant to be ascribed to Plato; I also accept uncritically the view that this position is appropriately ascribed to Plato. Both assumptions can be justified. See Burnet 1900, Introduction § 28, and Joachim 1951, 31. Woods (1982, 61) draws a similar conclusion regarding the parallel *EE* i. 8.

[2] I accept a version of Fine's (1984) treatment of Aristotelian separation, though I would want to offer a slightly different account of the modal force of 'substances have a capacity for independent existence'. For an alternative to Fine's approach see Spellman 1995. On Spellman's alternative, given in terms of numerical distinctness, see Chapter 6, n. 21.

(5) FOG causes good things to be good (*EN* 1096ᵃ34–ᵇ3), where this presumably requires no more than 'formal' causation.

The arguments Aristotle offers in *Nicomachean Ethics* i. 6 attack (1)–(5) variously. Not all of its arguments are clear, and not all of them focus crisply on any one of (1)–(5). That said, it is reasonably clear that Aristotle directs his arguments for non-univocity primarily against (1).

The outlines of Aristotle's general argument against (1) is straightforward. Several times in the chapter, Aristotle aligns the homonymy of goodness with the homonymy of being. He tries to infer the non-univocity of goodness from some features of categorial predication:

> Further, since the good is spoken of in as may ways as being is (*eti d' epei tagathon isachôs legetai tô(i) onti*)—for <it is spoken of> in <the category> of what-it-is <i.e. substance>, for example god and mind; in quality, the virtues; in quantity, a suitable amount; in relative, the useful; in time, the propitious; in place, a location; and in other <categories>, other such things—it is clear that the good cannot be something universal, common <to all good things>, and single. For if it were, it would not be spoken of in all the categories, but in one only. (*EN* 1096ᵃ23–9)

Very little in this argument is clear; nothing is uncontroversial.

The framework of the argument, at least, seems clear and familiar:

(1) If the good is spoken of in many ways, it is homonymous.

(2) The good is spoken of in many ways.

(3) Hence, it is homonymous.

(4) If the good is homonymous, it is non-univocal.

(5) If the good is non-univocal, it cannot be something universal, common [to all good things], and single.

(6) Hence, the good cannot be something universal, common [to all good things], and single.

Given the framework we have established, (1), (4), and (5) are uncontroversial. On the assumption that homonymy and multivocity are coextensive, (1) is unproblematic; (4) merely reflects that non-univocity is a necessary condition of homonymy; and (5) is analytic. Proposition (3) follows directly from (1) and (2), as does (6) from (3)–(5). Hence, the only premise requiring articulation and defence is (2).

Proposition (2) is supposed to follow from facts about the way goodness is predicated in different categories. Indeed, goodness, Aristotle maintains, is predicated in as many ways as being is (*EN* 1096ᵃ23–4).

The content of Aristotle's argument accordingly raises an important but as yet unanswerable question: in how many ways is being spoken?

More to the point, the success of this argument depends upon the distinctive thesis that being is homonymous and is predicated non-univocally across the categories. If being has many senses, and goodness has as many senses as being has, then goodness too has many senses. The non-univocity of being is thus held to be sufficient for the multiplicity of goodness.

This argument raises two simple questions. First, is being said in many ways? Second, why suppose that goodness is said in as many ways as being is? Why suppose, that is, that being and goodness march in step at all? This second question raises some awkward problems for Aristotle concerning the alleged categorial parallels between being and goodness. I will not, however, address them. For the first question already receives an unfavourable answer: Aristotle never provides a compelling argument for the homonymy of being.[3] Consequently, I do not regard this argument, in its present formulation, as successful.[4] Fortunately, Aristotle does not here argue that the non-univocity of being is necessary for the multiplicity of goodness. Accordingly, it remains possible that a distinct argument, or a reformation of this argument, will succeed.

Even so, the categorial argument raises important questions about Aristotle's approach to goodness. Consequently, it merits a fuller appraisal.

8.3 THE CATEGORIAL ARGUMENT: MULTIPLE SENSES AND COMMON UNIVERSALS

Aristotle's suggestion that 'goodness' is used differently across the categories invites us to consider a variety of sentences:

(1) Socrates is good.

(2) God is good.

(3) Courage is good.

(4) A moderate amount of exercise is good.

(5) Early 1989 was a good time for investing in the market.

(6) This is a good location for a picnic.

---

[3] I assess his arguments in the next chapter.

[4] Many have been more sympathetic to Aristotle's primary argument, though their formulations of it often have little in common with one another. For some more sympathetic treatments, see Urmson 1987, 22–4; Ackrill 1972; MacDonald 1989; Kosman 1968; Joachim 1951; Hardie 1980; Gauthier and Jolif 1970.

These are just the sorts of examples Aristotle mentions. To these we might add any number of others, based upon the diverse sorts of things we call good: burritos, blankets of snow, fuel mileage, fights, flights of fancy, views, acidity levels, lives, prose styles, attitudes, jobs, angles, prospects, performances, saw blades, economic trends, and a tenor's rendition of 'Ombra mai fu'. Surveying this list—or any of a countless number of similar lists—we might be inclined to agree with Aristotle straight away that 'the good cannot be something universal, common <to all good things>, and single' (*EN* 1096ª28).

This inclination derives from seeming differences in shallow meaning or shallow signification. When I say that ice cream and democratic forms of government are good, surely I mean something different in each case? Evidently, in the first case I mean that ice cream is 'pleasant to eat', whereas democratic forms of government are 'well suited to meet the needs of organized societies'. Using the simple paraphrase test of *Topics* i. 15 should, then, establish homonymy. Substituting the relevant definitions, we have for:

(7) Ice cream is good.

and

(8) Democratic forms of government are good.

first

(7*) Ice cream is pleasant to eat.

and

(8*) Democratic forms of government are well suited to meet the needs of organized societies.

The accounts of 'good' thus differ. As further evidence: (i) any attempt to paraphrase (7) along the lines of (8*) or (8) along the lines of (7*) yields gibberish; and (ii) tests of oppositions will reveal the relevantly different contrasts. Hence, it might be concluded, 'good' is non-univocal: there is no one thing, a universal, present in all cases of goodness.

A Platonist ought to reject this easy inference. First, she should point out that Aristotle's reasoning may be reconstructed as follows:[5]

---

[5] See MacDonald 1989, 153–9 for a discussion of problems with what he calls the 'multiple-senses interpretation of homonymy'. He contrast the multiple-senses interpretation (A) with the 'multiple-natures' interpretation (B):

    (A) (1) If good were a single common nature, 'good' would have only one sense.
        (2) 'Good' has more than one sense.
        (3) Good cannot be a single common nature.   (*cont.*)

(1) If the good were a single universal, equally present in all cases of goodness, then 'good' would have at most one sense.

(2) The standard non-univocity indicators of *Topics* i. 15 reveal that 'good' has more than one sense.

(3) Hence, the good cannot be a single universal, equally present in all cases of goodness.

Both premises are problematic. Premise (1) is a problem because we might attach distinct senses to words, even though there is one, higher-order univocal universal to which they correspond.[6] Premise (2) is a problem for two reasons: (i) Aristotle's standard non-univocity indicators extend for the most part only to shallow signification; and (ii) the tests of *Topics* i. 15 do not capture what is distinctive about Aristotle's treatment of the good in *Nicomachean Ethics* i. 6.

I consider the problems with (1) and (2) in the next two sections. My intent in these discussions is to bring into sharper focus the question of how Aristotle can establish the non-univocity of goodness.

8.4 THE CATEGORIAL ARGUMENT: CRITERIOLOGICAL AND FUNCTIONAL APPROACHES

One may readily agree that appeals to the shallow semantic differences which 'good' manifests in various applications ought not to suffice for establishing the non-univocity of goodness. Indeed, one may argue, the lack of success of any such argument goes some way towards showing that homonymy in Aristotle has little or nothing to do with the senses of words. Now, I have argued that homonymy has a fair amount to do with the senses of words, because it is a doctrine about meaning, where meaning includes, but is not exhausted by, shallow meaning, the kind of mean-

(B) (1) If good were a single common nature, it would be spoken of in one way only (it would not be homonymous).
(2) Good is homonymous.
(3) So good cannot be a single common nature.
As I point out in Chapter 3, I am in some ways sympathetic to the multiple-senses interpretation, though I equate senses with shallow meanings and regard them as sufficient only in uninteresting cases, including most (but not all) discrete homonyms. So I agree with MacDonald that (A 1), as it stands, should not be regarded as providing compelling evidence for the homonymy of 'good'. Unlike MacDonald, however, I do not regard the multiple-senses and multiple-natures interpretations of homonymy as in competition with one another.

[6] See 2. 3 for a discussion of the relation between first- and second-order univocity.

ing required for linguistic competence.[7] Still, I agree with the critic's posture in this case, since I agree that appeals to shallow meaning will not
offer much in terms of a philosophically interesting doctrine of the
homonymy of goodness. Surely, at any rate, no Platonist should be moved
to abandon her views about goodness on the basis of appeals to shallow
meaning.

Some approaches to establishing non-univocity have not obviously
relied on appeals to meaning, shallow or otherwise;[8] others have expressly
disavowed appeals to meaning altogether, favouring a 'multiple-natures'
approach to the homonymy of goodness.[9] These approaches merit discussion because they reasonably attempt to base Aristotle's commitment
to the non-univocity of goodness on firm philosophical foundations—
foundations which appeal neither to word sense nor to the doctrine of categories as such. In this sense, they attempt sympathetic reconstructions of
Aristotle's categorial argument which do not advert directly or immediately to the non-univocity of being. For this reason, they are to be preferred to attempts to interpret the categorial argument strictly in terms of
the homonymy of being. Unfortunately, as I shall argue, they do not succeed in finding an alternate route to the non-univocity of goodness.

In the ensuing discussion, I distinguish two versions of the categorial
argument, one *criteriological* and one *functional*. I also distinguish two
forms of the criteriological argument, one straightforwardly epistemological and the other broadly metaphysical.

The criteriological interpretation rests upon the thought that the 'criteria for commending different things as good are diverse and fall into
different categories; and this is enough to show that "good" does not stand
for some single common quality'.[10] According to this approach,
Aristotle's main point about the multiplicity of goodness does not derive
from predicating goodness of items in distinct categories. The subjects of
the predications are mostly irrelevant. Rather, distinct predications of
goodness of a subject (or subjects) could be grounded in different ways,
according to radically different criteria, where the criteria themselves fall
into distinct categories. Thus, consider:

(9) . . . is good *because* or *in virtue of its being* god.

(10) . . . is good *because* or *in virtue of its being* virtuous.

---

[7] See 3. 8–11.

[8] See Woods 1982, 70–4; Hardie 1980, 56–8; Kosman 1968; and especially Ackrill 1972.
I consider Ackrill's view in the text.

[9] I have in mind especially MacDonald (1989), who makes the best case for what I will
call the functional approach. I discuss MacDonald's view in the text.

[10] Ackrill 1972, 22.

(11) ... is good *because* or *in virtue of its being* timely.

(12) ... is good *because* or *in virtue of its being* useful.

Here the driving insight would be not that goodness is predicated of objects in distinct categories (though it may be) but that the grounds for predicating goodness in these distinct cases will be categorially diverse.[11]

This argument evidently rests upon the premise that if *F* is predicated of *a* and *b*, and the criteria in accordance with which it is predicated in these instances diverge, even to the point of belonging to different categories, then *F* is predicated non-univocally. Thus, if 'being a body' is predicated of *a* merely in virtue of the fact that *a* occupies space, but is predicated of *b* in virtue of the fact that *b* manifests psychic activities, then 'body' is predicated non-univocally.

This premise is false.[12] One might understand the notion of criteria in either of two ways, as an evidential notion or as a constitutive notion.[13] That is, one may conceive of a criterion as a ground for believing something, as evidence for its being the case that *p*; or one may construe a criterion as what it is for something to be an *F*, as, for example, a criterion of a horse's being a good racehorse is its being able to run quickly over a specifiable distance. On either conception, the criteriological argument fails, because categorially distinct criteria are compatible with univocity.

Consider first the criterion as an evidential notion. Perhaps some café dwellers in Vienna will surmise that *S* is an American on the basis of their observation of her. One offers as a criterion that *S* smells of soap, another than *S* dresses well, and yet another that *S* walks with an odd overconfidence. They do not regard 'American' as non-univocal. On the contrary, their accounts of what it is to be an American are in perfect harmony with one another. It is merely that the evidence they offer for regarding *S* as an American differ, even significantly. This is unproblematic. There may be distinct criteria for applying one and the same predicate. Hence, understood as an evidentiary notion, distinctness of criterion is not sufficient for non-univocity.

If we intend instead to advance a constitutive notion of being a criterion, then we have a broadly metaphysical rather than epistemological thesis in mind. Thus, for example, my criterion for regarding a play as 'amusing' might be its witty dialogue, its improbable situations, its farci-

---

[11] Ackrill credits this view to Pacius and Alexander, both of whom rely on the categorial argument when commenting on *Topics* 107ª3–12.

[12] My objection here echoes MacDonald's complaint against Ackrill. See MacDonald 1989, 158–60, esp. n. 26.

[13] Ackrill does not distinguish these different ways of something's being a criterion.

cal plot, or its ebullient acting. Indeed, I might say its being amusing just is its having one or more of these traits. But I would not therefore be entitled to infer that 'amusing' is non-univocal. Rather, I would accept these various traits, taken either individually or corporately, as constitutive of the play's being amusing. Its being amusing, however, need not differ from case to case. What is meant in each case in calling the play amusing is that it is entertaining in a pleasantly diverting sort of way. This suggests that different traits may constitute something's being *F* even while *F*-ness is itself a perfectly univocal concept. Consequently, even if we understand it as a non-evidentiary notion, distinctness of criterion is not sufficient for non-univocity.

Perhaps it will be countered that this response does not acknowledge the full force of the constitutive interpretation of the criteriological version of the categorial argument. It may be conceded that different states of affairs can constitute one and the same quality, even while the quality in question is perfectly unified. Even so, it is objected, an adequate counter-example to the constitutive variation of the criteriological version would need to show more than just this. In addition, it would need to show how a single notion can be constituted by different states of affairs, *where those states of affairs are themselves in different categories*. After all, the point about goodness was not merely that different states of affairs could constitute goodness in different contexts. On the contrary, the point was that states of affairs *in different categories* could constitute goodness, thereby being criteriological for goodness, with the result that no one quality could capture what it is to be good.

This rejoinder has some merit. First, it takes very seriously the thought that the argument in question is a categorial argument, rooted in irreducibly distinct sorts of beings. Second, it calls attention to the fact that constitution, understood as a non-evidentiary notion, should help ground non-univocity by making disparately good things have different real natures. If 'being solicitous' constitutes goodness in lackeys and 'providing instructions for the production of delicious meals' constitutes goodness in recipes, then perhaps we should refrain from thinking of goodness as a single common quality.

This response is best understood as insisting that the criteriological argument, understood constitutively, makes the claim that when two conditions are met, *F* cannot be predicated univocally of *a* and *b*. These conditions are: (i) the criteria in accordance with which *F* is predicated of *a* and *b* diverge, and (ii) the criteria themselves belong to distinct categories.

Understood in this way, the constitutive interpretation of the criteriological argument nevertheless fails. It may be dangerous to cross the street without first looking both ways, and it may be dangerous to expose oneself to ambient carcinogens, and it may also be dangerous to be in the wrong place at the wrong time. But being dangerous is not therefore non-univocal. On the contrary, what it is to be dangerous will remain fixed across these applications, even though what constitutes danger will diverge, and will diverge in such a way that what constitutes danger will derive from distinct Aristotelian categories. Consequently, the constitutive approach to the criteriological version of the categorial argument is uncompelling.

Thus far, then, we have the following result. The criteriological approach to the categorial argument must be understood in either evidentiary or constitutive terms. Neither yields a satisfactory interpretation of the criteriological interpretation, however, because neither forces non-univocity. Therefore, if it rests ultimately on criteriological considerations, the categorial argument is unsound.

Recognizing the failure of the criteriological formulation of the categorial argument, one may turn to a broader functional interpretation.[14] The functional interpretation situates Aristotle's categorial argument in the broader context of its appearance in the first book of the *Nicomachean Ethics*. Although *Nicomachean Ethics* i. 6 falls in some ways outside the main thread of argumentation in the first book (1096^{a}11–15, 1097^{a}13–24), the categorial argument is nevertheless articulated within the framework of that book, and its central terms are consequently coloured by the discussions of surrounding chapters.

In particular, immediately following the categorial argument of *Nicomachean Ethics* i. 6, Aristotle introduces and defends a functional account of goodness for human beings, according to which 'the good and <doing> well seem to <reside> in the function' (*en tô(i) ergô(i) dokei tagathon einai kai to eu*; *EN* 1097^{b}26–7).[15] On this approach, for functional kinds at any rate, goodness consists in the realization of one's fullest nature, which is itself determined in functional terms. Thus, as Aristotle suggests, the goodness of a flute player is understood in terms of the function of flute playing; so too with other functional kinds, which, Aristotle argues, extend to human beings as such (*EN* 1097^{a}11–12).[16] In line with this approach, Aristotle identifies the good with 'that for the sake of which' an *F* engages in the kinds of activities characteristic of *F*s as such. Thus, the good for generalship will be victory, for flute playing it will be playing well, and for carpentry it will be building well. This follows in part from the

---

[14] MacDonald (1989) advances this interpretation most clearly and forcefully.

[15] I understand *to eu* at *EN* 1097^{b}27 as an implicit articular infinitive, *to eu poiein*.

[16] Aristotle's claims here are difficult and disputed. For our purposes, it will not be necessary to move beyond a generic formulation of Aristotle's approach. For more detailed appraisals, see Joachim 1951, 48–52; Whiting 1988; and especially Kraut 1989, 237–53 and 312–22.

method Aristotle proposes for determining the function of *F*s: one needs to isolate what is peculiar (*idion*) to *F*s as *F*s (*EN* 1097^{b}33–4), and to determine whether what is thus isolated constitutes the essence of *F*s.[17] Having done so, it will be possible to establish the functionally defined good in terms of the characteristic activity of that kind.

With this in mind, it is possible to formulate a functional interpretation of the categorial argument:

(1) For a functional kind *F*, the good for *F*s consists in 'that for the sake of which' *F*s act.

(2) That for the sake of which the members of any functional kind *F* act will be the *end* for those *F*s.

(3) Hence, for a functional kind *F*, the good for *F*s consists in their end.

(4) The ends for different functional kinds necessarily differ.

(5) Hence, what goodness consists in for different functional kinds necessarily differs.[18]

(6) If (5), then goodness is non-univocal.

(7) Hence, goodness is non-univocal.

To this general argument for non-univocity, the functional interpretation of the categorial argument adds a new and distinctive premise:

(8) The categories themselves are functional kinds.

(9) Hence, the goods for the categories are necessarily distinct.

(10) Hence, goodness is said in as many ways as being is.

This argument is an improvement over the criteriological formulation of the categorial argument. Even so, I shall argue, it too is unsound—and unsound in an instructive way.

Although (7) is not established by (1)–(6), I shall grant it for the moment to focus on (8), the point at which the functional specification of

---

[17] For a discussion of this proposal, see Kraut 1979.

[18] MacDonald (1989, 167) identifies as key claims in this interpretation: (a) 'the notion of good is equivalent to the notion of an end', and (b) 'the thing, state, or activity which constitutes something's end is what it is for the thing in question to be complete, lacking in nothing, as a thing of its kind'. He then observes: 'It follows from (a) and (b) (plus the claim that the functions of things differing in kind are themselves different) that the real nature or property which constitutes the good for some particular kind of activity or thing will be different from the nature or property which constitutes the good for some different kind of activity or thing.'

the categorial argument is most distinctive. Proposition (8) has some points in its favour. Perhaps better than any other approach to the categorial argument, the functional interpretation, and (8) in particular, helps to make sense of some of Aristotle's perplexing examples. Thus, he claims that goodness is spoken of in the category 'of what-it-is <i.e. substance>, for example god and mind' (*EN* 1096ᵃ24–5), where these may be understood as especially good instances of the category in question. If one can first determine what the different categories are by assigning them various functions, then one can further determine whether individual members of that category fulfil that function well or poorly. The idea, then, is that god and mind fulfil whatever function substances as such fulfil better than other members of that category, for example trees and grubs.

What are the functions of substances as substances? On the functional interpretation, Aristotle's commitment to the independence of substance is a sort of function for substances.[19] God and mind, then, count as paradigmatic substances, because they exist independently most fully and completely; no other substance satisfies the independence criterion in the way they do.

There are, however, three problems with thinking this way. First, the suggestion that substances as such have functions, and that their function is to be identified in terms of their satisfying an independence criterion, is difficult to understand. It is true that a necessary condition of being a substance is being separate (*chôriston*), where this is plausibly understood in terms of having a capacity for independent existence. It is therefore also true that this criterion can be employed to distinguish substances from non-substances. It does not follow immediately, however, that substance is a functional kind; nor is it clear how such a claim is to be understood. Surely the mere recognition that substances are differentiated from other kinds of things by their fulfilling an invariable necessary condition will not suffice for ascribing them the function of fulfilling that condition. Perhaps this is true for manifest functional kinds; the issue at present, however, is just whether substance is itself a functional kind.

Moreover, and this is the second problem, even if this worry can be addressed, it is unlikely that the functional approach can be generalized. Recall some of Aristotle's other examples: 'in quality, the virtues; in quantity, a suitable amount; in relative, the useful; in time, the propitious; in place, a location' (*EN* 1096ᵃ25–7). It makes no ready sense to think of

---

[19] MacDonald 1989, 169: 'Existing independently can be construed as performing a function, analogous to the way in which building and playing the harp are functions.' On the independence of substances, see *Metaphysics* 1019ᵃ1–4, 1028ᵃ33–4, and 1071ᵃ1–2.

the propitious as the functionally best instance of time, or the suitable amount as the functionally best instance of quantity. Nor is it at all clear how the virtues should be thought to be the best qualities along some functionally specified axis. How, for example, is being courageous a functionally better quality than being red? It is not; and there is little point in trying to compare one quality with another in terms of an artificially functional account. The reasons for this motivate my final problem for the functional approach. In order for there to be intra-categorial functionally specified forms of goodness, there must first of all be inter-categorial functional differences such that we can distinguish one category of being in functional terms. But the ten categories of being have no functions as such: they are not functional kinds.[20] Hence, (8) is false. Consequently, the functional interpretation, although initially forceful, is unsound.

That said, it remains to consider (1)–(6), which already offer an important contribution by attempting to establish the non-univocity of goodness in general functional terms which do not yet appeal to the difficult, distinctive (8). The core idea of (1)–(6) is to identify the good of a functional kind with its end, and then to note that the ends for different functional kinds vary not only in range but in kind. Some ends are products and some are activities. Activities themselves vary, from building to contemplating to flute playing to cutting. But since these ends differ, the goods with which they are identified differ. So goodness will differ and hence will be non-univocal.

This argument fails to establish non-univocity. A parallel from philosophy of mind will be illustrative. Someone might argue that beliefs for cats, dogs, humans, and androids are distinct. Beliefs in androids are constituted by silicon stuff, in humans and other animals in neural states, but of different sorts in different species. So since these beliefs are different real states in these different types of entities, belief must be non-univocal.

The response is clear. Belief may be univocal, even if it is multiply realizable. After all, belief may be a functional kind itself: its nature is to be explicated by the inputs, outputs, and relations to other mental states it plays in systems of sufficient complexity. Oddly, in this sense, the functional interpretation of the categorial argument carries the seeds of its own undoing. A Platonist should simply respond that goodness may, like belief, be second order, though univocal. Although it is true that it may be realized differently by different functional kinds, goodness may yet admit of a perfectly univocal account.

In summary, then, the categorial argument is best understood in terms

---

[20] See 9. 11 for a discussion of this possibility.

of either the criteriological or the functional interpretation. The criteriological interpretation, taken either evidentially or constitutively, fails to
establish non-univocity. The functional argument, though more sophisticated, also fails. First, its striking attempt to treat the categories as functional kinds whose members have functionally specified ends cannot be
made to work. Even if one were to grant this contention in the case of substance—and I have argued that one should not—the point could not be
generalized, because the categories are not functional kinds. Moreover,
the functional interpretation, like the criteriological interpretation, cannot establish non-univocity. Put more polemically, the categorial argument cannot *force* non-univocity. And it is worth recalling the dialectical
situation at this point of the discussion. The Platonist may reasonably
demand that Aristotle actually force non-univocity, because Aristotle had
responded to the Platonist's univocal account not merely by disputing its
content but by rejecting its general form. Aristotle denies that goodness
'is something universal, common <to all good things>, and single' (*koinon
ti katholou kai hen*; *EN* 1096ª28), and so accepts a debt to establish its non-
univocity. The categorial argument fails to make good on this debt.

8. 5 FORCING NON-UNIVOCITY: A THOMISTIC PARALLEL

This problem with the categorial argument is a special case of a pervasive
problem with arguments for non-univocity. It is consequently worth
emphasizing this problem in some detail. We have seen that a Platonist will
sometimes reasonably respond to claims of non-univocity by conceding
that a term is not univocal at the first order, only to insist that there is a
second-order univocity which the critic has missed.[21] Thus, for example,
someone might be misled into believing that 'poison' is non-univocal,
because she had missed the second-order, causal account which was
wholly univocal. In the present instance, someone might respond to
Aristotle's initial argument by insisting that there is a univocal account of
goodness, for example, that *x* is good just in case *x answers to an interest*.[22]
A Platonist could then insist that this is a perfectly univocal account of
goodness, one which held true equally of democracies and ice cream.

It is unclear how Aristotle's framework of categories provides any
resources for responding to this argument. After all, his detractors will

---

[21] See 2. 3.
[22] This is what Ziff seems to maintain (1960, 176–82).

insist, they themselves accept the claim that goodness is spoken across the categories; but they will deny that this provides evidence for its non-univocity. They will, in short, treat cross-categorial predication as akin to multiple realizability, and they will rightly argue that neither suffices for non-univocity.

Aristotle needs a way to force univocity. His initial failure to do so can be appreciated by contrasting his approach to the non-univocity of goodness with that of Aquinas. For Aquinas, using premises alien to Aristotle, is able to force non-univocity—on the assumption, of course, that the alien premises are defensible. Aquinas's argument comes out in several passages, including:

> It is clear that nothing is predicated univocally of God and other things . . . for the things that God has made receive in a divided and particular way that which occurs in God in a simple and universal way. (*SCG* i. 32. 1–2)

> Whatever is predicated of many things univocally is either a genus, a species, a differentia, an accident, or a proprium. But, as we have demonstrated, nothing is predicated of God as a genus or a differentia; hence, neither is anything predicated as a definition, nor as a species (which is constituted of a genus and differentia). Nor, as we have shown, can there be any accident in God, and therefore nothing is predicated of him either as an accident or as a proprium, since proprium belongs to the genus of accident. It follows, then, that nothing is predicated univocally of God and other things. (*SCG* i. 32. 4)

It is noteworthy in these arguments that Aquinas holds that if the premises are true, it is *not possible* for goodness to be predicated univocally of God and creatures.

Simplified and streamlined for the present purposes, Aquinas's argument is:[23]

---

[23] Aquinas's arguments are actually articulated more minutely than this streamlined presentation suggests. He offers four distinct arguments:

(i) Ways of Predicating (*SCG* i. 32. 2):

(1) If *F*-ness is predicated of God, *F*-ness is predicated in a simple and universal way.
(2) If *F*-ness is predicated of creatures, it is predicated in a divided and particular way.
(3) If *F*-ness is predicated in diverse ways (that is, here simply, there in a divided way), then *F*-ness is predicated non-univocally.
(4) Hence, if *F*-ness is predicated of God and creatures, *F*-ness is predicated non-univocally.

(ii) Subjects of Predication and Their Modes of Being (*SCG* i. 33. 3):

(1) Suppose *F*-ness is exactly the same in each of its instances.
(2) *F*-ness would then be predicated univocally only if it were received 'according to the same mode of being'.
(3) God shares his mode of being with no creature.
(4) Therefore, no predicate is predicated univocally of God and creatures.   (*cont.*)

(1) God is good.

(2) God is intelligent.

(3) Professor Peabody is good.

(4) Professor Peabody is intelligent.

(5) God is absolutely simple.

(6) Therefore, God has no parts.

(7) Therefore, goodness and intelligence in God are *identical*.

(8) Goodness and intelligence are not identical in Professor Peabody.

(9) Therefore, by the identity of indiscernibles, the goodness which Professor Peabody manifests cannot be identical with God's goodness.

(10) Therefore, goodness is not predicated univocally of God and Professor Peabody.

At this point, Aquinas takes himself to have established non-univocity. He adds:

(11) The goodness of God and Professor Peabody are not altogether distinct.

and infers:

(12) Therefore, goodness is predicated analogously (= homonymously) of God and Professor Peabody.

The argument clearly has a number of troublesome premises. My concern is not with their truth but with their form.

Of special interest is Aquinas's strategy in implicitly appealing to the

---

(iii) Participation (*SCG* i. 32. 6):
(1) If *F*-ness is predicated univocally of *a*, *b*, and *c*, then *a*, *b*, and *c* are *F* by participation in *F*-ness.
(2) God is never *F* by participation in *F*-ness.
(3) Therefore, if God and creatures are *F*, *F*-ness is not predicated univocally.

(iv) Priority and Posteriority (*SCG* i. 32. 7):
(1) If *F*-ness is predicated of diverse subjects according to priority and posteriority, then *F*-ness is not predicated univocally.
(2) If *F*-ness is predicated of God and creatures, then *F*-ness is predicated with reference to some priority and posteriority.
(3) Therefore, if God and creatures are *F*, *F*-ness is not predicated univocally.

identity of indiscernibles. The property 'goodness', as it occurs in the case of God, itself has a property which the goodness of the creature Professor Peabody lacks. God's goodness has the property of being identical with wisdom; Professor Peabody's goodness is not identical with wisdom. God's absolute simplicity requires lack of division and, hence, identity of all of God's attributes. Professor Peabody is not simple. Nothing compels the identification of his goodness with his intelligence.

Of course, when faced with this Thomistic argument, one might legitimately question God's absolute simplicity, on the grounds that it leads to a contradiction: the property of goodness both is and is not identical with intelligence. Aquinas's response, in brief, is to deny that we have grounds for a contradiction because the goodness of Professor Peabody and the goodness of God are distinct. This is, after all, why he treats goodness as non-univocal. What appears contradictory to some provides a principle in accordance with which Aquinas thinks he can force non-univocity, namely, the identity of indiscernibles as applied to goodness in distinct contexts.

Accordingly, the question for Aristotle, as he responds to the proponent of the univocity of goodness, is whether he can appeal to a similar principle of differentiation to force non-univocity. His bare appeal to the doctrine of categories did not suffice.

8.6 FINAL GOODS, INSTRUMENTAL GOODS

After Aristotle's main argument is complete, he considers a dialectical response, precisely of the form we might well expect from the Platonists. Aristotle's response, I shall argue, provides a defensible way of arguing for non-univocity.

This is fortunate, since, as we have seen, Aristotle's initial argument ties the homonymy of goodness to the doctrine of categories in a way which does not immediately capture its non-univocity. It attempts to establish a distinct form of goodness for each of the ten categories of being. In response to a possible Platonist response to this initial, categorial argument, Aristotle expressly limits the range of applications, thus committing himself to homonymy across a much narrower range than we have so far considered:

[A] certain objection to what was said may be made clear because <the Platonists> were not speaking of arguments concerning every good and because <only those> good things pursued and loved intrinsically, by themselves, are spoken of in accordance with one form, while those productive of these or somehow preservative of them or destructive of their opposites are spoken of <as good>

because of these and in another way. It is clear, then, that good things would be spoken of in two ways, some intrinsically, and the others because of these. (*EN* 1096ᵇ8–14)

Aristotle entertains the following rejoinder to his categorial argument. The Platonist is imagined as responding that the hypothesized Form of Goodness does not cover all cases of goodness; nor does it correspond to the general term 'goodness' in all of its applications. Rather, the Platonist claims, the Form of Goodness is realized by all and only things which are good in themselves, as ends, which are loved and pursued for their own sakes.

So far, Aristotle has no response. It is only because he ascribes to the Platonist an additional thesis that he can respond as he does. The Platonist adversary is made to hold that there are also other goods 'spoken of <as good> because of these and in another way' (*dia tauta legesthai kai tropon allon*; *EN* 1096ᵇ12–13). Having secured this premise, Aristotle can infer directly that, if so, 'good things would be spoken of in two ways' (*dittôs legoit' an tagatha*; *EN* 1096ᵇ13). Crucially, in this dialectical context at any rate, Aristotle adjusts his estimation of the number of ways in which good is spoken of downwards from many ways, as many as being (*EN* 1096ᵃ23–4) and in all of the (ten) categories (*pasais tais katêgoriais*; *EN* 1096ᵃ28–9), to two ways (*dittôs*). The dialectic of the passage is a bit unclear. It is not, for example, at all clear that Aristotle wishes to grant the Platonist the proposed restriction. Rather, Aristotle seems to reason that *even if* we grant the Platonist the opportunity to restrict the Form of Goodness to intrinsic goods, any admission of non-intrinsic goods would already force non-univocity. At the same time, Aristotle subsequently seems to abandon one natural consequence of the categorial argument: that goodness, like being, is core-dependent (*EN* 1096ᵇ26–31). This may suggest second thoughts about the force of the categorial argument on Aristotle's part; and these second thoughts may be occasioned by the Platonist rejoinder he considers.

However that may be, Aristotle's response to the Platonists provides a principle of differentiation lacking in his main categorial argument. Now Aristotle, no less than Aquinas, can appeal to the identity of indiscernibles in arguing for non-univocity. Moreover, he can do so without any appeal to the difficult, idiosyncratic premises to which Aquinas appeals. In short, he can employ Aquinas's strategy without relying on Aquinas's premises.

Aristotle's argument in response to the Platonist rejoinder can be expanded in either of two ways.

(A) (1) Some things are good in themselves; others are good only in so far as they are conducive towards these intrinsic goods.

(2) Things good in themselves are essentially good.

(3) If x is essentially F, then x is necessarily F.

(4) Hence, whatever is good in itself is necessarily good.

(5) Things which are good only instrumentally are not essentially good.

(6) If x is not essentially good, x is necessarily good only if goodness is a logical or categorial property.[24]

(7) Goodness is not a logical or categorial property.

(8) Hence, things which are only instrumentally good are contingently, and not necessarily, good.

(9) If x is necessarily F and y is contingently F, then x and y are F in different ways.

(10) Hence, things which are good in themselves and instrumental goods are good in distinct ways.

(11) Hence, goodness is not predicated univocally of them.

This first expansion appeals to different ways of having the predicate goodness and thus invites, especially, questions about (A 9) as a generator of non-univocity.

There is something plausible about Aristotle's suggestion, thus expanded, but so far it is not as crisp as Aquinas's way of forcing non-univocity. Proposition (A 9) seeks to derive non-univocity from the fact that some good things are essentially good, while others are merely contingently good. One might wonder, however, whether this is sufficient for invoking Aquinas's strategy of appealing to the identity of indiscernibles. According to Aquinas, in one instance goodness has the property of being identical with intelligence, while in another instance it lacks this property. Aristotle, by contrast, has only the premise that goodness is sometimes predicated necessarily, while at other times it is predicated contingently. This does not yet seem directly analogous. If oddness is predicated necessarily of the number three but only contingently of the number of apples

---

[24] This premise is required to rule out the suggestion that some things might be necessarily good without being essentially good. Thus, perhaps everything is necessarily identical with the number seven or not identical with the number seven. But it is not part of Socrates' essence that he is either identical with the number seven or not identical with the number seven. Aristotle treats essences as explanatorily basic. See *Topics* 102ª18–24 and 128ᵇ34–6, as well as Chapter 3, n. 16.

on my table, this in no ways compels us to agree that 'oddness' is non-univocal across the applications.

This objection has only a spurious plausibility. Taken *de re*, the number of apples on my table, say three, is not contingent, even if, taken *de dicto*, the proposition 'the number of apples on my table is odd' is contingent. The reason oddness is univocal in both applications is that we are thinking in both cases *de re*, and not *de dicto*. For it is only when we think *de dicto* that we have reason to suppose that oddness is predicated contingently. But oddness is not predicated contingently; rather, the proposition about the apples on my table being odd is contingently true. Consequently, we have not been given a reason to doubt (A 9).

We can see that (A 9) ultimately proves unproblematic in another way, by expanding Aristotle's argument somewhat differently:

(B) (1) Some things are good in themselves, others in so far as they are conducive towards the good.

(2) The goodness of things good in themselves is complete, final, and lacking in nothing.

(3) The goodness of things not good in themselves is not complete, final, and lacking in nothing.

(4) Hence, the goodness of things good in themselves has a property which the goodness of things not good in themselves lacks.

(5) Hence, by the identity of indiscernibles, the goodness of things good in themselves is not the same as the goodness of things not good in themselves.

(6) Hence, goodness is not predicated univocally of them.

This, as Aristotle says, is why goodness is spoken of in two ways, and why goodness is said of instrumental goods 'in another way' (*allon tropon*; *EN* 1096^{b}13).

Aristotle has an argument for distinguishing two senses of goodness which is distinct from his earlier categorial argument. It is an argument which succeeds against his Platonist opponent, but only on the condition that the opponent grants a distinction between things good in themselves and things good only instrumentally. To the extent that this concession is granted, Aristotle's argument has some force. The concession is not a difficult one; it is not unfamiliar in the Platonic corpus (e.g. *Rep.* 357^{b}2–358^{a}2); and it is one not easily denied. If one insists, for example, by way of arguing for higher-order univocity, that '*x* is good' means '*x* answers to an interest', one merely postpones the question. For some things answer to an interest because they satisfy that interest in a primary and complete way, while other things answer to an interest only to the degree that they provide a pathway to what satisfies an interest in a primary and complete way.

To this extent, then, Aristotle is justified in arguing for the non-univocity of goodness. He could also have exploited a second principle, namely FD, the functional determination thesis. For we saw that FD parcels entities into kinds but that Aristotle appropriately denied that all kinds are functional kinds.[25] Because human beings have functions (*EN* 1097^{b}22–7, 1098^{a}8–12), their goodness can be functionally specified. This will be impossible for kinds which lack functions, e.g. rain. Still, rain can be good, bad, or indifferent. In so far as it can be good, rain has a goodness which is unlike the goodness attaching to functionally defined entities. Goodness in those cases will consist in the realization of a functionally specified goal. Here too, then, goodness appears non-univocal: there is a functional and a non-functional notion, and there appears to be no second-order univocal account which stitches them together.

This argument for non-univocity also adopts the strategy implicit in Aquinas's argument. That is, it appeals to the fact that the goodness in one case has a property lacked by the goodness of another case. In one case goodness has the property of being functionally determined; for non-functional kinds, this is not possible. Hence, by the identity of indiscernibles, goodness is non-univocal.

In embracing a broadly Thomistic strategy for establishing non-univocity, Aristotle has a better chance of showing why the Platonist cannot return the discussion to a stalemate simply by appealing to the possibility of a higher-order univocal account of goodness. It should be allowed, however, that this is not a strategy which Aristotle articulates in any express way when recounting methods for establishing non-univocity, as he does, for example, in *Topics* i. 15.[26] Still, it is plausibly regarded as an implicit regulative principle for those methods; it is a strategy evidently employed in *Nicomachean Ethics* i. 6 itself, in the context of addressing reasonable Platonic concerns.

8. 7 CONCLUSIONS

Most explications of the homonymy of goodness in Aristotle rightly focus on what I have called his categorial argument. This is, after all, his princi-

---

[25] See 1. 5.   [26] On *Topics* i. 15, see 2. 4.

pal argument for attempting to establish the non-univocity of goodness. In this argument, Aristotle seeks to infer, in an unclear way, the homonymy of goodness from the doctrine of categories and the multivocity of being. Because in its most promising formulations the categorial argument rests upon criteriological and functional principles insufficient for generating non-univocity, I have not found in this argument any clear or compelling case for non-univocity. Further, if we revert to its simplest formulation, according to which it appeals to the homonymy of being for one of its premises, the categorial argument fares no better. For if the arguments of the next chapter are correct, this premise is false.

Fortunately, Aristotle does not rely on just that one argument for non-univocity. Instead, when considering a dialectical objection to his primary, categorial argument, he offers a second, less ambitious argument. This second argument is much more promising. It implicitly relies on the strategy for forcing non-univocity later made explicit by Aquinas, who equally wished to establish the non-univocity of goodness. In employing this strategy, Aristotle, unlike Aquinas, adverts only to the features of good things which a Platonist could and should accept, namely that some good things are intrinsically good, while others are good only in so far as they are conducive towards those primary goods. With this premise, Aristotle can exploit the strategy upon which Aquinas relies to force non-univocity. Of course, non-univocity is not sufficient for association; I do not find any clear case for the association of goodness in accordance with CDH_4 in *Nicomachean Ethics* i. 6.

With this partial defence of Aristotle's treatment of the homonymy of goodness, I turn to Aristotle's most celebrated appeal to homonymy, the homonymy of being. In this case, I cannot offer even a partial defence.

9

The Homonymy of Being

9.1 A FRAMEWORK FOR DISCUSSION

When he sought to explicate and defend Aristotle's conception of the homonymy of being in his seminal *Von der mannigfachen Bedeutung des Seienden nach Aristoteles*,[1] Franz Brentano quite reasonably focused initially on two passages of the *Metaphysics*. In the first passage, Aristotle claims:

Being is spoken of in many ways, but with respect to one source. For some things are called beings because they are substances; others <are called beings> because they are attributes of substance, others because each is a route toward substance: either destructions or privations or qualities or productive or generative of substance; <still others are called beings because they are> things spoken of in relation to substance, or negations of one of these or of substance. For this reason we say that even non-being *is* a non-being. (*Met.* 1003ᵇ6–10)

In the second, Aristotle advances an evidently non-equivalent, more encompassing contention:

Being, spoken of simply, is spoken of in many ways, one of which was[2] the accidental, another was the true (with non-being as the false), and beyond these there are the schemes of the categories (e.g. what <something is>, quality, quantity, place, time, and if <being> signifies something else in this sort of way); and further beyond all these as in potentiality and actuality. (*Met.* 1026ᵃ33–ᵇ2)

Brentano thought these passages came to the same thing,[3] in as much as each carries a commitment to the same doctrine (cf. also *Met.* 1003ᵃ33–4, 1017ᵃ8–ᵇ9, 1028ᵃ10–31, 1051ᵃ34–ᵇ6). From each he elicited a fourfold homonymy of being: (i) accidental being as opposed to being in itself (*on kata sumbebêkos* as opposed to *on kath' hauto*); (ii) being as truth (*on hôs alêthes*); (iii) categorial being; and (iv) being in potentiality as opposed to being in actuality (*on dunamei* as opposed to *on energeia(i)*). He

---

[1] Brentano 1862/1975.
[2] Aristotle here refers back to the discussion of the mulitivocity of being in *Metaphysics* v. 7.
[3] Brentano 1862/1975, 3–5.

consequently structured his entire dissertation on the homonymy of being around these four senses.

It is doubtful that the passages Brentano cites reduce to just these four senses of being,[4] or even that they make substantially the same claims about being. Even so, Brentano's scheme of classification brings out an essential point about the homonymy of being: that Aristotle's doctrine admits of weaker and stronger formulations. The weaker formulation holds merely that there are various uses of 'is' we would want to recognize as distinct, and that the semantic function of 'is' depends in part on the circumstances of its use. For example, 'is' has both veridical and existential functions, roughly Brentano's (ii) and (iii). Thus, there is a difference between saying 'There really was a Marquis de Sade' (= existential) and 'The Corsican Napoleon Bonaparte was never a Frenchman, not the way the Marquis de Sade was' (= veridical).[5] The second statement encourages the paraphrase, 'The Marquis de Sade was a genuine Frenchman, but Napoleon was not' or 'The Marquis de Sade was a true Frenchman, whereas Napoleon was not'. The first does not admit of any such paraphrase.

If his view of being comes only to this, Aristotle's conception of the homonymy of being is defensible. It is also philosophically bland. In particular, it will do nothing to support the distinctively Aristotelian claim that 'being' is used differently across the categories, as claimed in Brentano's (iii), categorial being. On the contrary, in terms of this weak doctrine, all the members of the distinct categories may be said to be in the same sense, because they all exist. Although he never promotes it as such, Brentano's (iii) contains a much stronger doctrine, one not on par with his (i), (ii), and (iv). In each of those cases, some single sense of being is exhibited. By contrast, (iii) promises a further multiplicity of being in terms of the categories themselves. According to (iii), 'is' occurs with a unique, discernible sense in each of the ten categories. For this reason, his (iii) should not be regarded as one sense of being, alongside three other equivalent senses; for it contains itself a stronger, much more controversial doctrine about the homonymy of being which no simple appeal to semantic intuition can sustain.

Put another way, we might have sought a reasonable, relatively accessible doctrine in Aristotle by regarding the homonymy of being as a cer-

---

[4] Mansion (1976, 218–44) recognizes three of Brentano's four senses of being, because she collapses his (i) and (iii). Aubenque (1962, 163–79) concentrates on (i) and (iv), on the grounds that these are the two required for Aristotle's purposes, namely the refutation of sophistical Megarian puzzles.

[5] Mansion (1976, 235–7) offers this sort of interpretation. She concludes (243): 'L'être au sens de vrai n'est pas non plus un genre d'êtres: il n'est que l'expression de la fidélité de l'esprit au réel.'

tain sort of semantic doctrine, answering to the familiar point that 'is' is used in several ways, and therefore has several senses: as a copula ('The boy is obstreperous'); as expressing identity ('George Eliot is Mary Anne Evans'); as an existential quantifier ('There is an x such that x is bald and the king of France'); and, in its participle form, 'being', as a substantive ('Parsimony cuts two ways: it pertains equally to the number of beings postulated and to the number of the types of beings postulated'). It should be clear already that Aristotle is concerned not with the general question about the general semantic functions of 'is', but rather with the question of whether one sense of 'is'—presumably the existential sense or predicative sense—admits of yet further multiplicity, a veiled sort not as readily detectable as the general semantic and syntactic multiplicity attaching to 'is'. As Owen says: 'At various places Aristotle says things which show how the verb "to be" in its existential rôle or rôles can have many senses.'[6] Perhaps Aristotle can show how being is homonymous, even when thinking only about being in its existential sense. This is the project of trying to establish the stronger doctrine of the homonymy of being.[7] This is also the project I consider and assess in this chapter.

The over-arching argument of the chapter is straightforward. As we have seen, establishing core-dependent homonymy involves a three-stage process: (i) one must demonstrate non-univocity; (ii) one must then establish association; and finally (iii) one must establish core-dependence in line with CDH_4.[8] Most commentators, fastening on the intuitively compelling idea that substance, *ousia*, is a primary form of being, assume non-univocity and move directly to the core-dependence of non-substantial being on substantial being.[9] In proceeding this way, they assume something

[6] Owen 1965*a*, 264.

[7] Aristotle never says, 'Being is homonymous'. Still, he frequently claims that being is spoken of in many ways (*pollachôs legomenon*); and since Aristotle treats homonyms and multi-vocals as co-extensive (see 1. 3), his commitment to the multivocity of being is sufficient for its homonymy.

[8] On these aspects of core-dependence, see especially 4. 5.

[9] A typical example comes from Code (n.d., 8): 'Every non-substantial being is a being by virtue of bearing the right kind of relation to the primary beings, to substances. There is no single condition by virtue of which "things that are" are properly called "things that are". Some things are so-called in a primary way, others in a derivative way.' Similarly, Code claims (10): 'In its other uses the predicate "being" signifies either "what X is like", or "how much X is", or each of the other things predicable in the way these latter things are. All other things are beings derivatively. Anything that is a "being", but not in the way substances are, is properly called a "being" by virtue of standing in some appropriate relation of ontological dependence to something that is a "being" in the primary way. Some things are beings because they are qualities of substances; and so on. As required by the account of "being" given in [Metaphysics iv. 2], the term "being" applies primarily and without qualification to substances, and derivatively to all else. This requires that substances are the primary beings, and hence that substantial being (that is, their being, or their *ousia*) is the

false. The arguments of Aristotle's commentators consequently fail to establish non-univocity; and Aristotle's own arguments lapse into internal inconsistency. Hence, since establishing non-univocity is a necessary condition of establishing core-dependent homonymy, no defensible account of the homonymy of being emerges from Aristotle's writings.

9. 2 ARISTOTLE'S DEVELOPMENT ABOUT BEING

After introducing the science of being qua being in *Metaphysics* iv, Aristotle denies that being is homonymous, where he clearly intends to deny only that being is a discrete homonym:[10] 'Being is spoken of in many ways, but with respect to one thing and a single nature and not homonymously' (*Met.* 1003ª33–5). This has suggested to some that sometime between the *Organon* and *Metaphysics* iv, Aristotle made a realization about being which induced him to accredit a science whose credentials he had earlier called into question. Owen thought this realization was that beings, like other homonyms, could be focally connected to a core meaning which would provide the form of unity required for scientific autonomy and integrity.[11] On his approach, the core meaning of *to on* would be *ousia*, and this is the gloss he provides for Aristotle's contention that the science of being qua being is a science with the structural characteristics of a special science even though it is universal:

In sum, the argument of *Metaphysics* IV, VI seems to record a new departure. It proclaims that 'being' should never have been assimilated to simple cases of ambiguity, and consequently that the old objection to any general metaphysics of being fails. The new treatment of *to on* and other cognate expressions as *pros hen kai mian tina phusin legomena*, 'said relative to one thing and a single character'—or as I shall henceforth say, as having *focal meaning*—has enabled Aristotle to convert a special science of substance into the universal science of being, 'universal just inasmuch as it is primary'. (*Met.* 1003ª23–4, 1026ª30–1; cf. 1064ᵇ13–14)[12]

Owen's developmental hypothesis encodes a series of philosophical questions: (1) What does Aristotle come to appreciate about being that forces a change in his thinking? (2) What is it for beings to have focal meaning, or any form of focal association, for beings to be, in our terminology, core-dependent? (3) Is there some defensible doctrine of core-dependent

---

primary kind of being. Indeed, for something to be a primary being, or a substance, is for it to possess the primary kind of being, or *ousia*.' In a different way, and in a different context, Frede (1987) makes the same sort of assumption. I consider Frede's proposal below in 9. 5.

[10] On discrete versus associated homonyms, see 1. 2.

[11] Owen 1965*a*.   [12] Owen 1960, 184.

homonymy underpinning both Aristotle's recognition of a single science of being qua being *and* his criticisms of Academic theorizing?

These questions are prompted by Aristotle's introduction of a science of being qua being in *Metaphysics* iv. 4:

There is a science which studies being qua being and the things belonging to it in itself. This is in no way the same as any of the sciences called departmental [or: called the special sciences]. For none of the others investigates universally about being qua being, but rather each cuts off a certain part of it (i.e. of being) and studies it <as it> coincides, as e.g. among the sciences mathematics <does>. Since we are seeking the principles (*archas*) and the highest causes (*aitias*), it is clear that these must be the <principles and causes> of some nature in itself. If, then, those seeking the elements of being were also seeking the <highest> principles, it would be necessary <for these elements> to belong to being not coincidentally, but qua being. Hence, it is also necessary for us to find the first causes of being qua being. (*Met.* 1003ᵃ21–32)

Among the surprising claims here are, first of all, that there is a science of being qua being, but also that this science is universal and not compartmentalized. Here Aristotle seems to have changed his mind about the possibility of such a science and so to have developed a new insight about being, about science, or about both.[13]

If this introduction represents a change in Aristotle's thinking, what has prompted it? As we have seen, Owen ascribes it to a discovery about homonymy. Earlier, Aristotle failed to appreciate the core-dependent character of the homonymy of being; when he now uncovers it, Aristotle sees that the science in question is possible after all.

Of course, Aristotle never announces that his discovery into the core-dependent homonymy of being underwrites the possibility of a science of being qua being. Hence, any reason Owen or anyone else might offer must be garnered from other things Aristotle says about being, the putative science of being, or core-dependent homonymy.[14]

---

[13] Owen (1965a, 259) thinks Aristotle must have changed his mind about being, since his introduction of a science of being qua being positively contradicts his earlier denials of the existence of a science of being: 'There is a contradiction between these claims which . . . is central to Aristotle's philosophical development.'

[14] Owen himself tries to establish Aristotle's discovery indirectly, suggesting that this is the best explanation for his structuring faulty anti-Platonic dilemmas whose solution lay precisely in terms of the discovery that being qualifies as a core-dependent homonym. As we have seen, because Aristotle had clearly come upon the notion of core-dependent homonymy by the time he structured those very dilemmas, the argument cannot be that Aristotle simply did not have the notion available to him. It must be, rather, that Aristotle had not yet seen how his doctrine of core-dependent homonymy could be *applied* to being. Owen (1960, 184) claims that at the time of writing the *Eudemian Ethics*, Aristotle had 'not yet seen its [i.e. focal meaning's] application to such wholly general expressions as "being" or "good"'.

If we knew that Aristotle failed to appreciate the phenomenon of core-dependent homonymy at the time he denied the existence of a science of being in his earlier work, including, for example, the *Eudemian Ethics*, but that he had discovered it at the time of his postulation of a science of being qua being in *Metaphysics* iv. 4, then we would have some evidence, however inconclusive, that his discovery about the core-dependent character of the homonymy of being induced him to revise his thinking about the possibility of there being such a science. Given that we do not know this, we need to see what might have prompted him to apply it to his analysis of being only after having written the *Organon* and *Eudemian Ethics*. Then we would have another sort of evidence, still less conclusive, that he connects core-dependent homonymy with being in a way which reveals how the proscribed science is after all possible.

Even this sort of evidence is hard to produce. Aristotle clearly denies the existence of a science of being (*EE* 1217^{b}25–35; cf. *An. Post.* 92^{b}14; *Top.* 121^{a}16–19, 121^{b}7–9). He also seems to claim that there is no sense of 'being' such that everything in every science has being, even if the entities presumed to exist in each of the special sciences can be said to be in an analogical sense relative to one another (*An. Post.* 76^{a}37–40, 88^{a}36–^{b}9). This claim seems extreme and impossible to defend. Just as it is difficult to appreciate why the principle of non-contradiction should have more than one meaning, one in biology and another in politics, it is difficult to see why anyone should want to hold that the entities studied by astronomy and biology cannot be said to exist in precisely the same sense, but only by analogy. That the principle of non-contradiction has distinct ranges of application in distinct spheres of inquiry is evident enough. But this hardly entails that the principles employed in these distinct disciplines are themselves distinct, related only because the work they do in structuring those sciences is analogous. In any case, these sorts of remarks pepper the *Organon* and may be thought to show that Aristotle must not yet appreciate the core-dependent homonymy of being.[15] Since he does appreciate it later, it is possible that just this discovery forces his change of mind.

Nothing Aristotle says in the *Organon* is incompatible with a core-dependent analysis of being.[16] He evidently embraces one part of such a

[15] So Owen 1960, 189.
[16] Hence, I disagree with Aubenque (1962, 181): 'On ne peut s'empêcher alors de remarquer que la doctrine des catégories est ici invoquée pour appuyer une démonstration exactement contraire de celles que nous trouvions, au sujet du bein dans les *Topiques*, *l'Éthique à Eudème* et *l'Éthique à Nicomaque*. Dans ces derniers textes, il s'agissait de montrer qu'il n'y a pas une science une du Bien, parce que le bien se dit en autant de sens différents que l'être. Ici, au contraire, il s'agit d'établir qu'il y a une science une de l'Un, parce que l'Un comporte autant d'espèces que l'être et que les espèces de l'Un correspondent à celles de l'être.

claim by denying there is one sense of being appropriate for all beings when he insists that being is not a genus. When he further claims that being is analogous across the special sciences, he evidently sees some connection. Although this connection falls far short of any commitment to homonymy, it is not incompatible with an analysis in terms of core-dependence. If this is correct, we have no specific evidence tying the *Organon* to a rejection of core-dependent homonymy for being. Given that we have clear evidence revealing Aristotle's acceptance of core-dependent homonymy as a general phenomenon in the earlier works, including the *Organon*, and given that we have no specific evidence that he refused to regard being as a core-dependent homonym, we have no reason at all to suppose that Aristotle at any stage of his career refused to regard being as a core-dependent homonym. His silence on the issue is simply silence.

Still, it might be countered, *something* must explain Aristotle's change of mind about the possibility of a science of being. And if this is not an insight into the character of being, what could it be? And if it is an insight into the character of being to the effect that being is somehow more than analogically related across the special sciences, what could it be except the realization that it is a core-dependent homonym, like health and medicine? Surely this is the reason Aristotle compares being to health and medicine almost immediately after introducing the science of being in *Metaphysics* iv. 4?

These questions have a false presupposition. For they presuppose that Aristotle *did* change his mind about the possibility of a science of being. It is noteworthy that when commenting on Aristotle's conception of the analogical character of being in the *Organon*, Owen observes that this 'does nothing to show the possibility of a general science of "being and the necessary characteristics of being", which takes the common axioms of the sciences as part of its subject-matter just because those axioms hold good of being *qua* being'.[17] He is correct, but only because Aristotle's conception of being as analogical is perfectly neutral with respect to core-dependent homonymy. What is noteworthy, however, is Owen's quoting Aristotle as introducing *a general science of being* in *Metaphysics* iv. 1. Aristotle does not introduce such a science; rather, he announces a science of *being qua being*, where this is not clearly the science whose existence he

Or, il n'est pas douteux que "espèces de l'être" de la *Métaphysique* ne désignent pas autre chose que les significations de l'être des *Topiques* et des deux *Éthiques*: le parallélisme même des problèmes montre que, dans les deux cas, il s'agit bien des catégories. La contradiction entre les deux séries de textes est donc flagrante.'

[17] Owen 1960, 189.

had earlier denied. For the 'qua' locution delimits the scope of the inquiry in a way fully compatible with Aristotle's earlier denial of a science of being.[18]

Indeed, if Aristotle had earlier denied the possibility of a science of being in part because he denied the existence of a genus of being, then if the science he now announces is the precise science whose existence he had earlier denied, we should expect him now to accept the existence of such a genus. That is, if we understand him in the *Organon* as holding that there is a science of being only if there is a genus of being, and as denying the existence of the science because he denies the existence of the genus, then when he changes his mind, we should expect him to change his mind about the very science whose existence required a genus of being. There is no evidence of his postulating any such genus in the *Metaphysics*. On the contrary, in so far as he insists that the being under investigation in the science of being qua being is spoken of in many ways (*Met.* 1003[a]33), and so homonymous, Aristotle evidently continues to hold that there is no genus of being.

Of course, Aristotle may have changed his mind about whether a science of being required the postulation of a genus of being. Here again we have no evidence, because Aristotle does not explicitly sever the connection between the science of being and a genus of being. For, again, the science introduced in *Metaphysics* iv. 1 is the science of being qua being.

I do not mean in this discussion to deny that Aristotle developed in his thinking about the possibility of a science of being. Much less do I want to claim that Aristotle's views on these matters were static, spanning the earlier *Organon* and the more mature *Metaphysics*.[19] On the contrary, I am confident that his introduction of a science of being qua being in *Metaphysics* iv. 1 represents a kind of development for Aristotle.[20] Rather, I hope to have established only the more modest point, that Aristotle's introduction of the science of being qua being need not have stemmed from an insight into the core-dependent character of being. If this is correct, then there is no need to regard his putative discovery of the core-dependence of being as the crucial insight which paves the way for a science of being qua being.

---

[18] Bolton argues for a similar conclusion, but along slightly different lines (1995, 423–4): 'In the *Metaphysics* Aristotle continues to maintain his doctrine of categories, on which there are many different ways of being not subsumable under any generic real kind *being* (see Z. 1. 1028[a]10 ff.). So the *Metaphysics* does not reject the doctrine of the *Analytics* that being is not a genus and that there is, thus, no science that has *being* for its subject genus.'

[19] On dating, see n. 1 to Chapter 1.

[20] Roughly, Aristotle has come to appreciate the falsity of premise B in the argument offered in 2. 9.

9. 3 FIVE APPROACHES TO THE HOMONYMY
OF BEING

Aristotle need not rely on an insight about the core-dependent character of the homonymy of being to legitimate a science of being which would otherwise be untenable for him. This is fortunate since, as I will argue, there is no workable analysis of being as a core-dependent homonym. Now, Aristotle does think of being as a core-dependent homonym. Hence, if the arguments of the remainder of this chapter succeed, his views in this regard are indefensible. The effects of the arguments of the last section will, however, localize the damage. For if Aristotle required a commitment to the core-dependent homonymy of being in order make the science of being qua being possible, then that entire science would be impossible, in much the way that the science of being *tout court* was held to be impossible in the *Organon*. Conditions necessary for the possibility of the proposed sciences would fail to obtain, with the result that this science could never get off the ground.

The argument of the remainder of this chapter is, then, critical. It proceeds in two stages. The vast bulk is given over to an argument by enumeration. I argue that five distinct approaches to the homonymy of being fail to reveal any defensible doctrine. Like any argument by enumeration, this argument is compelling only to the degree that it can claim to be exhaustive. I do not claim complete exhaustiveness on its behalf, if this is to be understood deductively, with general reasons given for supposing that these five are the only possible approaches. Still, if it addresses every plausible approach available and succeeds in revealing serious problems for each, the first phase of the argument can claim to be complete in the sense that it refutes all promising reasons anyone might have for accepting the thesis it assails.[21] In any case, the second stage attempts to offer a

---

[21] I do not discuss two noteworthy contributions. First, Bostock (1994, 45–52 *inter alia*) identifies two distinct formulations of the multivocity of being, both of which merit attention. According to the first, 'the "is" which connects a substance as a subject with a non-substance as predicate, i.e. the "is" of coincidental predication, introduces a different relation for each different category of predicate' (51–2). According to the second, being applies differently to coincidental items and items in their own right, and 'the being of such things is different in different cases, for what being is for a thing is given in full by its full definition, and this of course is different in different cases' (52). The second line of reasoning, which he regards as 'confused' (52), has the advantage of lining up the number of kinds of being with the number of categories but affords no priority to substance. The first line affords some priority to substance, but it 'does not work out too well' (48), in part because it gives us little reason to suppose that being is said differently across the categories. Because these alternatives come out in the context of a commentary, Bostock does not develop them; it is clear, however, that he is right to question whether either is promising. A second noteworthy discussion occurs in Matthews 1995. Matthews holds that being is, for Aristotle, syncategorematic. His suggestion is intriguing, but I do not think he shows that this is

more general, more speculative argument diagnosing a problem with the homonymy of being as Aristotle conceives it. I intend this second argument to be comprehensive in a way that my argument by enumeration is not.

Most of the approaches in the first phase of the argument have been adopted in one form or another by Aristotle's expositors. Others I develop from textual suggestions, in light of the general framework of homonymy developed in Part I of this study. I begin with two distinct proposals by Owen, whose work contains a serious effort to provide a philosophically defensible interpretation.

The five proposals I consider are these: (i) the homonymy of being is a thesis about reduction or translation; (ii) the homonymy of being is a thesis about distinct ways that 'is' ascribes predicates to subjects; (iii) the homonymy of being results from Aristotle's grasping that different kinds of being manifest different ways of being; (iv) the non-univocity of being can be grounded, like some other equivocals, in FD, the thesis of functional determination; and, most important, (v) the homonymy of being can be inferred directly or indirectly from the theory of categories. This last approach is most important because it represents what I take to be Aristotle's own most sustained attempt to establish the non-univocity of being.

Some of these approaches seek to ground or justify the homonymy of being. Others merely try to explicate it. In different ways, each tries to provide a satisfactory account by indicating why Aristotle might adopt the thesis as true. My arguments against these approaches are mixed, contending in some cases that the views represented are not plausibly Aristotelian and in other cases that whatever their pedigree, the views recounted are indefensible. Mainly I am interested in a critical enterprise: I aim to establish the evaluative claim that none of these approaches articulates and defends a successful argument for the non-univocity of being.

9. 4 DEFINITION, TRANSLATION, AND FOCAL CONNECTION

Owen offers different accounts of the homonymy of being in different papers and sometimes offers non-equivalent accounts within an indi-

so. Indeed, if he were right, it is hard to see how being could be any sort of core-dependent homonym, since it would be unlike health precisely because it would have 'no independent sense at all' (236). That said, Matthews poses just the right question for anyone who accepts the homonymy of being (233): ' "'To be' is said in many ways" is one of Aristotle's favourite sayings. But how many is "many"? In exactly how many ways did Aristotle think "to be" is said?' Answering this simple question proves surprisingly difficult.

vidual paper. In an earlier paper, 'Logic and Metaphysics in Some Earlier Works of Aristotle',[22] Owen explicates the homonymy of being by appealing to notions of reducibility and translatability. Some statements are statements explicitly about substances, e.g. 'Socrates is a man'. Other statements seem to be about non-substances, e.g. 'Triangles have three sides', or 'Orange is closer to red than it is to purple'. Since beings are core-dependent homonyms, there should be a core way of being, or a core sense of 'being', and all other beings, or senses of 'being', should depend on the core case. If there is a core sense of being upon which all non-core senses depend, then, Owen plausibly suggests this sense will be *ousia*. But what will be the form of dependency Aristotle seeks? Because he thinks that homonymy is a semantic (or metaphysically rich semantic) notion,[23] Owen naturally talks in terms of reduction or translatability: 'The claim that "being" is an expression with focal meaning is a claim that statements about non-substances can be reduced to—translated into—statements about substances.'[24] If this is correct, any statement whatsoever should be statable in terms of substances, since everything is either a substance or a non-substance.

Owen's contention is at once too weak and much too strong. It is too weak because it does not make explicit that translations could be one-way only: the reducibility in question captures the primacy of *ousia* only if there is some asymmetry in translation. Otherwise, we could equally translate talk about substances into talk about non-substances, and we would then overlook the core-dependent character of the homonymy in question, since all derived homonyms are allegedly referred to some one source (*mia archê*). Owen's constraint is also too weak because it does not make clear the role of definition in clear cases of homonymy. It is not only that non-core homonyms *can* be translated into talk about their core instances. Rather, their definitions *must* make appeal to the definitions of their core

---

[22] Owen 1960.

[23] Owen talks indifferently about words and things as homonyms, even though he specifies at one point that Aristotle regards things as homonyms. He quite regularly talks in terms of the senses or meanings of homonyms, where this can only be understood as a broadly semantic interpretation. As one of many examples: 'Where he might be expected to say that all the subordinate senses of *on*, "being", must be defined in terms of a primary sense of that expression, what he says is that all senses of *on* must be defined in terms of *ousia*, "substance", just as all senses of "healthy" must in terms of "health": a formulation which makes no provision for the *priority* of one sense of *on*. But he then talks as though he had provided for that priority; and the explanation is plain—*on* in its primary sense is *ousia*' (1957, 184 n. 16). Owen's talk of sense here commits him to a semantic interpretation, as does his habitual preference for mentioning rather than using the words or expressions to be defined.

[24] Owen 1957, 192.

instances. 'Healthy' is homonymous because any account of, say, a healthy complexion must, in order to be complete, make appeal to the account of 'healthy' as it occurs in 'Manfred is healthy' or, more cumbersomely, 'Manfred has health'. Put more formally, Owen's constraint is too weak just because it does not respect the formal condition on any adequate account of core-dependent homonymy,[25] which all of Aristotle's examples from the *Topics* onwards presuppose, namely, CDH_4. As we have seen, according to this condition, healthy complexions and healthy regimens will be associated because definitions of 'healthy' in these cases will equally make appeal to some core notion of being healthy, by being indicative or productive of health.

If Owen's account is too weak in these ways, it is simultaneously much too strong in others. Reflecting on Aristotle's views about the ontic dependency relations between non-substances and substances, mooted first and most baldly in *Categories* 5, we might be inclined to think concerning the *Metaphysics* as well that if there were no *ousiai* there would be nothing else. In the *Categories*, Aristotle simply asserts that 'if there were no primary substances, it would be impossible for any of the other things to exist' (*mê ousôn tôn prôtôn ousiôn adunaton tôn allôn ti einai*; 2ᵇ6ᵇ⁻ᶜ). Although this claim admits of weaker and stronger interpretations, Aristotle here asserts minimally that a necessary condition of the existence of any non-substance is the existence of a substance. Hence, all non-substances are dependent for their existence on the existence of primary substances. We might in turn wed this notion of ontic dependency (called 'natural priority' by Owen) to a thoroughly semantic notion of core-dependent homonymy, yielding the claim that all talk of derived cases can be reduced or translated into talk about core cases. Since in the case of being the core is *ousia*, this would yield precisely Owen's account.

We can see that this is much too strong when we reflect on the very examples of core-dependent homonymy Aristotle uses to illustrate the homonymy of being in *Metaphysics* iv. 2, namely, health and medicine. Healthy complexions are indicative of health; healthy regimens are productive of health. Standing in the relations of 'producing' and 'being indicative of' are in no way sufficient to guarantee translatability or reducibility. Indeed, if we accept FCCP,[26] then we can conclude directly that merely standing in one of the four causal relations will not suffice to ensure translatability. This is at any rate clear from the examples Aristotle develops in *Metaphysics* xi: 'A treatise and a razor [or a scalpel, *machairion*] are called "medical" because the one arises from medicine and

---

[25] See 4. 2, 4. 3, and 4. 5 for a discussion of this formal condition.
[26] On FCCP, see 4. 4.

the other is useful for medicine' (*Met.* 1061ᵃ3–5). Of course a scalpel is useful for cutting; but nothing suggests that every statement about scalpels can be reduced to some statement about medicine. One might say: 'The scalpel is made of good metal', or 'The scalpel used to belong to my father, whose father was a doctor'. As these examples illustrate, although any account of what a scalpel is will make appeal to medicine, no stronger requirements are necessary or desirable. So too for beings: 'being' may be homonymous even though not all statements about all beings are reducible or translatable into statements about *ousiai*. Owen's demand adds too much to CDH₄. It requires not only that accounts of non-substances make reference to substance but that such accounts must be specifiable purely in terms of substances. CDH₄ itself makes no such demand. It falls far short of any commitment to reduction or translation, however weakly construed; and Owen nowhere justifies the very strong addition he proposes.

9. 5 BEING* AND BEING**

Though they speak to the issue of non-univocity, Owen's remarks about translatability seem mainly concerned with the nature of the dependency relation non-core instances of being should bear to core instances. Elsewhere, Owen focuses more narrowly on non-univocity when he offers a second, non-equivalent, explication and defence of the non-univocity of being.[27] This second approach does not rely on any commitment to translation or reduction. Instead, Owen claims to find in Aristotle an insight into two forms of being, forms whose discernment will escape all but the most skilled at detecting non-univocity.

Owen begins with the observation, made by others as well,[28] that Aristotle holds that everything which exists falls under some sortal or other. Everything of which it is true to say that it exists is such that there is some predicate, other than existence, which can also be predicated of it. This is no doubt, for example, part of what Aristotle means when he claims that 'for living things, to be is to be alive' (*to de zên tois zôsi to einai estin*; *An.* 415ᵇ13; cf. also *GC* 318ᵇ25; *EN* 1168ᵃ4–5; *Top.* 148ᵃ26–ᵇ3).[29] So

---

[27] Owen 1965*a*, 264.

[28] Geach 1954–5 and Anscombe and Geach 1961.

[29] The context here is worth bearing in mind. Aristotle is seeking to show that since the soul is an *archê* and an *aition*, and these admit of different kinds, the soul must be an *archê* which is an *ousia*; for it is the kind of *aition* which explains the being of something (*An.* 415ᵇ8–14).

for everything which is, there is some property F (where 'F' signifies some property other than existence) such that it is F.

This much is apt. Now, however, Owen seeks to tease out of this simple insight something arresting. Aristotle's remark about life counts as one of the 'places where Aristotle says things which show how the verb "to be" in its existential rôle or rôles can have many senses'.[30] For something to be a child is not at all what it is for something to be a glimmer in a parent's eye. Hence, being for a child is not the same as being for a glimmer in an eye. When we see that a exists only because a is an F, and b exists only because b is a G, we should see also that what it is for a to exist is not the same thing as it is for b to exist. To grasp this point is to grasp the non-univocity of being.

What is the point to be grasped here? We may agree that being a child is not at all the same sort of thing as being a glimmer in a parental eye. This is because 'being a child' is not at all the same property as 'being a glimmer in a parental eye'. Perhaps we may also want to concede that nothing could be a glimmer in an eye unless it is the glimmer in the eye of a substance, whereas something can be a child in its own right. Yet neither of these concessions requires the further point Owen introduces on Aristotle's behalf: 'Aristotle wants to dispel the myth that there is equally something in common to sharks and shyness on the plea that each of them is a *being* or *existent* or *thing* of some kind.'[31]

There is a slip here. It may be true that sharks and shyness are not things of some one kind. Yet their belonging to different kinds does not entail that they do not equally exist. Of course, Owen's point may simply trade upon the more familiar claim that being is not a genus. If so, it collapses into a plea for homonymy which I consider and reject below.[32] If this is not the point, then Owen has nowhere exposed a myth for Aristotle to dispel.

Still, Owen rightly draws attention to a passage of *Metaphysics* viii. 2 which provides an argument for the homonymy of being we have not yet encountered. Aristotle claims:

Some things are spoken of by the composition of their matter, just as those things like honey-water which are <spoken of> by mixture;[33] other things <are spoken of> by being bound, e.g. a bundle; others by being glued, e.g. a book; others by being nailed, e.g. a casket; others by more than one of these; still others by position, e.g. the threshold and the lintel (for these differ by being placed in certain

---

[30] Owen 1965*a*, 264.   [31] Ibid. 265.

[32] See 9. 8.

[33] Aristotle typically opposes *krasis* and *sunthesis*, although here the former is offered as an instance of the latter (*GC* 328ᵃ8). Perhaps here he intends them in a non-technical sense, as at *De Anima* 407ᵇ30.

ways); others by time, e.g. dinner and breakfast; others by place, e.g. the winds; others by the affections of objects of sensation, e.g. by hardness or softness, density and rarity, and dryness and wetness; and some things <are spoken of> by means of some of these, and other things by means of them all; and generally, some by excess and some by defect. So, it is clear also that 'is' is spoken of in just as many ways. For a threshold is whatever lies in a certain way, and being for a threshold[34] signifies something's lying in this way, and being for ice[35] <signifies> being solidified in a certain way. (*Met.* 1042^{b}16–28)

Here Aristotle makes no obvious appeal to other, more familiar motivations for regarding beings as homonymous. In place of arguments appealing to being *kath' hauto* or to there being no genus of being, Aristotle forges a link between composition and locations and their modes of being.

Owen argues that Aristotle's argument here 'has a claim to be heard'.[36] This is correct, but not for the reasons Owen provides. He focuses especially on Aristotle's closing remarks and argues that the negations of various claims to existence yield different paraphrases. In line with a test provided in the *Topics*,[37] this reveals non-univocity in the original claims. The idea is this. Consider 'lights' in (a) and (b):

    (a) She lights the way with a hand-held torch.

and

    (b) He lights his pipe with a mother-of-pearl lighter.

The opposite of 'lights' in (a) is 'fails to illuminate', whereas in (b) it is 'fails to ignite'. So Aristotle will rightly infer that 'lights' in (a) and (b) is homonymous.

Owen attempts to apply the same test to 'exists' in (c) and (d):

    (c) The ice exists.

and

    (d) Socrates exists.

The denial of (c) is 'the ice has melted', whereas the denial of (d) is 'Socrates has died'. This reveals that 'exists' is non-univocal in (c) and (d). In (c) 'is' means 'is solid', whereas in (d) it means 'lives'. Since being solid is not the same thing as living, 'exists' is homonymous.

There are several problems with this proposal. First, it is far too

---

[34] Reading *oudô(i)*, on the strength of Alexander's paraphrase.

[35] Reading *to krustallô(i) einai* with Bonitz (1848–9) and Jaeger (1957) on the strength of Alexander's paraphrase, as against Ross (1924, ii), who reads *to kruistallon einai* at 1042^{b}27–8, on the strength of E and J.

[36] Owen 1965*a*, 266.
[37] *Top.* 106^{a}12–22. See I. 4.

profligate. If every time we denied the existence of something by characterizing the conditions under which it ceased to exist, 'exist' would be not
merely homonymous but wildly and uncontrollably so.[38] Socrates ceases
to live; ice melts; mud dries; fallen trees rot; lights are extinguished; hearts
stop pumping; crackers crumble; wine turns to vinegar; vegetables go to
seed; and the wind stops blowing. Applying Owen's test, we should
uncover a new form of existence opposing every possible way of perishing. Surely Aristotle cannot hold that 'exist' is so thoroughly homonymous. As he himself rightly claims, 'If one were to say that <a word>
signified an unlimited number of things, clearly no reasoning would be
possible; for not to signify one thing is to signify nothing' (*Met.* 1006^{b}5–7).

Noting this sort of problem, Owen proposes that 'Aristotle's answer is
the theory of categories'.[39] The theory of the categories places an upper
restriction on the modes of existence and so on the number of ways being
can be homonymous. If there are ten categories, then the number of ways
being is homonymous cannot multiply beyond ten. Yet the number of
ways things can perish far exceeds these ten. Surely Owen is not entitled
to the assumption that the number of ways things can perish serendipit-
ously equals the number of categories. Hence, either the test Owen
employs fails to establish non-univocity or the categories place no
restriction on the ways being can be homonymous. If the test he employs
fails to establish non-univocity, then Owen has failed to establish the
homonymy of being in the way proposed. If his test succeeds, then the
homonymy of being has run afoul of the wholly reasonable constraint
that being cannot be homonymous in an indefinitely large number of
ways.

The right inference to draw is that we do not establish the homonymy
of being by the periphrastic test of opposites. This, then, is the second
problem with Owen's proposal: that distinct entities require distinct
modes of composition does not show that they exist in different senses.[40]

---

[38] Matthews (1995, 234) appreciates this consequence: 'Thus, on Owen's reading, the
unsettling claim [that "to be" is said in an indefinitely large number of ways] comes to this:
"to be" is said in as many ways as there are kinds of things to talk about.' He infers, however that on this reading ' "There exist frogs and toads" would be guaranteed to be false'.
This inference need not follow, however, as long as we allow 'exists' to sustain two interpretations, as we do for 'in' in the sentence: 'He rode into town in a new car and a foul
mood.'

[39] Owen 1965a, 265.

[40] Owen (1965a, 271) is aware of this problem: 'Other criticisms cut deeper. Grant that
it is a different thing for a man to exist and for a sandal to exist; why should this imply that
"exist" has different senses, any more than "work" must be taken to have different senses
because it is one thing for a banker to do his work and another for a hangman to do his?
Great numbers of words (it will be argued) are in one way or another specially dependent
on the context for their particular force.' Owen does not respond to the objection, com-

Books are glued and become unglued, whereas twigs are bundled and become undone. But they both exist before falling apart. Moreover, if we think Aristotle's point depends essentially on the fact that it is a different thing for a book to exist and for a bundle of twigs to exist, as Owen does,[41] then we risk treating clear synonyms as homonyms. It is in some sense a different thing for a tiger to be an animal and for a butterfly to be an animal; but they are nevertheless synonymously animals (*Cat.* 1ᵃ8). One may reasonably wish to deny that it is a different thing for each of these distinct kinds of animals to be an animal. This is because 'animal' is synonymous. It is analogously a reasonable thing to deny that it is a different thing for a book and a bundle of twigs to exist.

Owen is therefore wrong to exploit Aristotle's test of opposites in arguing for the non-univocity of being. He is not therefore wrong to appeal to principles of negation in general. Although he does not introduce it as such, Owen provides an independent motivation for the homonymy of being which may not succumb to the objections so far registered. Although introducing his distinction as one between senses of 'existence', Owen shifts to speaking about a use of 'is' 'not in any sense a predicative use of the verb but a use which is parasitic upon all predicates',[42] and finally, after introducing a technical distinction, into claiming, 'To predicate being*, in the appropriate sense of the word, of men or icebergs, is tacitly to presuppose that there are** men and icebergs for this predicate to apply to'.[43]

The technical distinction is this. Existence** corresponds to the existential quantifier; existence* corresponds to a free-standing sense of 'is' which requires no further completion.[44] While it is true that everything which is* is some *F* or other, it is quite possible to say, truly, '*a* is no more', without specifying in addition the *F* which *a* no longer is. By contrast, if we wish to deny an existentially quantified singular proposition, we must minimally say: $\sim\exists x(Fx)$. So, Owen claims, there are two notions of existence: one which is always parasitic on some predicate, and one which is not. From Owen's perspective, in isolating being*, Aristotle has identified a sense of being whose context sensitivity reveals non-univocity. A lintel is* because it is positioned in a certain way. If it were otherwise placed, it would be a threshold. To say that a lintel is* no more, we need only recognize this fact.

menting, 'This was a hare to be raised, not chased now'. The issue is not merely one of context dependence, however. 'Work' is not context sensitive in the way that 'large' or 'heavy' or 'flat' are.

[41] Owen 1965*a*, 277.   [42] Ibid. 271.

[43] Ibid. 272.

[44] Owen is not always as clear about this distinction as one would like. This seems, however, to be the general purport of his view.

It is hard to appreciate Owen's argument here. If we agree that he thinks that everything which exists is some *F* or other, Aristotle presumably thinks that whenever we say '*a* is no more', we implicitly hold that there is some *F* which *a* no longer is. In these cases, *F* will be a substance sortal specifying the essence of *a*. If so, it is difficult to glean the distinction between two existential senses of 'is'. On the contrary, Owen's distinction between existence* and existence** seems to collapse into a more famil-iar distinction between existential and predicative senses of the verb. If so, then no strong form of homonymy has been established. It is merely the weak thesis of homonymy set aside at the beginning of this chapter, and by Owen himself in the beginning of his discussion.

These criticisms call into question Owen's defence of the homonymy of being. His interpretation of *Metaphysics* 1042ᵇ16–28 does not turn up any defensible grounding for this application of homonymy. Even so, he is right to focus on this passage. For there is another, more abstract way to characterize the argument it contains; and the argument so characterized is not one we have seen before. The argument Aristotle offers does not pro-ceed by inferring distinct modes of being from the conditions of dissolu-tion for entities which exist. Rather, Aristotle appeals to two features of things which exist: how they are composed and how they are placed. In this second connection in particular, he appeals to the conditions of signification, and he attempts to show how 'being' signifies different things relative to the context of something's existence.

Something may be described intrinsically as a piece of wood of certain type with certain dimensions, solidity, and weight. Any such description that lacks all reference to the wood's function or location, or indeed to any relational properties at all, will never suffice to characterize that piece of wood as a threshold or a lintel. Similarly, characterizing a wind in terms of its velocity and other intrinsic features does not yet determine whether the wind is the North Wind or the South Wind. So, we might infer, Aristotle wants to argue that since certain things are relationally specified, what it is for them to exist differs relative to their environments. But this cannot be quite his point, since he equally talks about the being of ice, whose characterization can presumably be wholly intrinsically specified. So, although part of Aristotle's concern, the context sensitivity of certain beings does not exhaust the point he seeks to make about being.

Clearly, and unsurprisingly, Aristotle's point involves conditions of signification.[45] He seeks to infer the homonymy of being from the fact that being for *F* signifies one thing, whereas being for *G* signifies another (*Met.*

---

[45] On signification and its role in establishing homonymy, see 3. 1 and 3. 3.

1042^{b}26–8). Some things are the same when intrinsically described, but distinct when positioned differently (lintels and thresholds, the winds); other things are what they are only when certain intrinsic, often compositional conditions obtain (books, bundles, ice). The 'is' in 'Ice is' signifies that there is something in a certain condition. It signifies that water is solidified. If the water were not so solidified, 'Ice is' would signify something false, or nothing at all. The 'is' in 'A lintel is' signifies that some material object is positioned in a certain way.

The homonymy of being, then, is revealed by—and indeed consists in—a being's signifying different real properties in different situations. The property 'being solidified', already restricted in its range of possible subjects, is not the same as the property 'being positioned above the entrance'. Since it signifies these distinct properties, 'is' is non-univocal.

This approach to the homonymy of being is the most substantive and successful of those canvassed. It has the distinct advantage of treating the homonymy of being as a special case of homonymy: like all homonyms, 'being' is homonymous in so far as it signifies different things in different applications. If it signifies a multiplicity of real properties, 'being' satisfies the first-stage demand for establishing core-dependent homonymy, since it turns out that it is after all non-univocal.

Even so, two problems remain. First, the problem of profligacy re-emerges, but at a different level. If a sufficient condition for establishing the non-univocity of being is its signifying different things in different contexts, and a sufficient condition for its signifying different things in different context is its picking out distinct contextual requirements for the existence of distinct things which are otherwise intrinsically the same, then there will be ungovernably many senses of 'being'. Note that the argument in question places no categorial constraints. Although genuinely distinct, the different contexts implicitly captured by 'is' in the various examples Aristotle provides do not respect any categorial limits. On the contrary, such contextual differences stem from the conditions of composition and relational states required for judgements of existence. These far exceed the ten categories, as Aristotle's open-ended list of cases shows.

Second, this argument is in the end a *non sequitur*, if it is supposed to establish the strong thesis of homonymy delineated at the beginning of this chapter. It infers from the *conditions* of something's existing to the *character* of its existence. A necessary condition of ice's existing is water's being solidified. A necessary condition of some wood's being a lintel is its being positioned above a door. That these conditions must be met does not yet show that the entities in question manifest different properties signified by 'exists', or that 'exists' is homonymous as it occurs in 'Ice exists' and 'A lintel exists'.

Here again, if the argument proved anything about the non-univocity of 'is', it would prove far too much. As it occurs in 'Socrates is an animal' and 'Felix the cat is an animal', 'animal' is synonymous. Appreciably distinct conditions must be met for Felix and Socrates to exist. For Socrates to exist, an organic body of a certain sort displaying certain characteristics must exist. For Felix the cat to exist, a body of a different sort displaying distinct characteristics must exist. More important, Socrates would not exist without his having intellectual capacities of a quite determinative sort, while Felix need have no such capacities. So distinct necessary conditions must be met for these different animals to exist. Even so, Aristotle treats 'animal' in these occurrences as synonymous (*Cat.* 1ª8). Consequently, he does not accept, as sufficient for the non-univocity of *F*, mere differences in the conditions requisite for something's existing as an *F*. Synonymous *F*s may have distinct conditions for their existing as *F*s.

Aristotle cannot, then, endorse an argument for homonymy based upon mere distinctions in preconditions of existence. As a special case, he cannot endorse such arguments for the non-univocity of being. Hence, in so far as it relies on any such argument, *Metaphysics* viii. 2 fails to establish the homonymy of being.

9.6 KINDS OF BEINGS, WAYS OF BEING

We see, then, that neither of Owen's approaches to the homonymy of being captures any plausible Aristotelian doctrine. This should induce us to start anew, from a different angle. Without first being mired in developmental questions of inter-scholastic dialectic, we might have focused directly on a long-standing problem internal to Aristotle's metaphysical programme, in an effort to trace its ramifications for homonymy. This problem concerns the relation between Aristotle's general and special metaphysics, that is between his introduction of a science of being qua being in *Metaphysics* iv and his subsequent development of a theory of substance, *ousia*, including especially his treatment in theology of the divine substance.[46] This problem is relevant to our discussion of the homonymy of being because one approach to its solution appeals crucially to a construal of the homonymy of being according to which different kinds of substances evince different ways of being. On this approach, the divine being exists focally as a core, while all other substances and non-substances exist somehow non-focally or derivatively.[47]

---

[46] See Shields 1987 for a brief overview of some of the main alternatives.
[47] Frede 1987 and Patzig 1960.

I do not here undertake a general treatment of the relation between general and special metaphysics. I am interested, rather, in a single formulation of the problem and the role homonymy is held to play in its solution. The formulation is this. Aristotle introduces a general science of being qua being in *Metaphysics* iv, and in subsequent books he pursues this science by investigating first natural substances, and ultimately one substance, god. It is unclear how this more special investigation, conducted most intensively in *Metaphysics* vi–viii and xii, constitutes the perfectly general enterprise introduced in *Metaphysics* iv. The general science of being qua being should be universal (*Met.* 1025^{b}9–10), whereas first philosophy concerns only one entity, the divine being, and so hardly seems universal. Aristotle himself does not seem to see any great problem here; he mainly comments, rather tersely, that first philosophy 'is universal in this way, that it is first' (*Met.* 1026^{a}30–1).

Some of Aristotle's expositors have treated the problem fully, in some cases by offering developmental accounts and in some cases by offering philosophical reconciliations.[48] Of special relevance to our present concern is Frede's approach, because it attempts to exploit the homonymy of being in its solution to the problem. Frede argues in three stages: (i) that theology deals not only with a special kind of being, god, but also more broadly 'with a particular kind or way of being, a way of being peculiar to divine substance';[49] (ii) that this way of being is explanatorily primary, so that accounts of all other ways of being must appeal to it if they are to be correct and complete; and (iii) since theology studies this focal way of being, one naturally and appropriately pursues the study of being as such by considering this core case.[50] Just as one who would wish to study everything pertaining to health would appropriately begin by considering what health itself is, so one investigating all forms of being should begin by considering exemplary being, as manifested by the divine substance.

This promising approach to the reconciliation of special and general metaphysics regards being as a core-dependent homonym and so treats being as meeting the primary conditions for any such account: non-univocity and core-dependence. Being displays non-univocity, because there are different ways of being; these ways of being are nevertheless ordered around some primary way of being, so that accounts of the derivative or non-core cases must asymmetrically appeal to being in the

---

[48] See Jaeger 1948, 226.

[49] Frede 1987, 84.

[50] Frede (1987, 84–5) adds that metaphysics equally concerns itself with universal principles (e.g. the principle of non-contradiction), and he suggests that these are germane especially to theology, since 'it will fall to the theologian to introduce them in an appropriate manner'. This additional consideration makes no appeal to core-dependent homonymy.

primary way. The homonymy of being, then, provides the key to solving a long-standing puzzle regarding Aristotle's metaphysical methodology. Since there are ways of being, one of which is primary relative to the others, a universal science of being qua being legitimately—indeed, necessarily—proceeds by analysing the core notion.

This account of the relation between general and special metaphysics is defensible only if its analysis of being as a core-dependent homonym succeeds. It is reasonable to assume that if distinct ways of being can be articulated, accounts of all non-core cases will need to refer back to the core, paradigmatic case of self-sufficient and complete being enjoyed by the divine substance alone. Still, this reasonable assumption is contingent upon there being multiple ways of being.

Frede's attempted reconciliation founders on just this point: because distinct senses or ways of being are not satisfactorily discriminated, the logically subsequent task of relating non-core cases to a primary case cannot be undertaken. Consequently, the core-dependent character of being cannot be exploited in effecting a reconciliation between general and special metaphysics.

Frede provides only an ostensive approach to the distinction between kinds and ways of being:

[L]et us try to understand how it is that theology is not concerned only with a particular kind of beings, but with a particular way of being, peculiar to its objects, and how it addresses itself to this way of being. By distinguishing a kind of beings and a way of being I mean to make a distinction of the following sort. Horses are a kind of beings, and camels are a different kind of beings, but neither horses nor camels have a distinctive way of being, peculiar to them; they both have the way of natural substances, as opposed to, e.g., numbers which have the way of magnitudes, or qualities which have yet a different way of being. The way magnitudes can be said to be is different from the way qualities or natural substances can be said to be. The claim, then, is that the way separate substances can be said to be is peculiar to separate substances.[51]

The idea is that some kinds of things, horses and camels, have a way of being in common; the way of being they have, however, is not the way of being other sorts of beings have. Camels and horses are natural substances and so have the way of being appropriate to all and only natural substances. Other sorts of substances are non-natural substances. These include the imperishable heavenly bodies and the imperishable, immaterial divine being (*Met.* 1071^{b}2–22). For expository convenience, let us concentrate on the non-natural substances I will call 'angels'. Angels are non-natural because they exist as substances without meeting the con-

---

[51] Frede 1987, 85.

ditions Aristotle sets for all natural substances: that they be realized in matter and have accidental properties.[52]

The question, then, concerns why one should suppose that natural substances and angels manifest different ways of being. That they are different kinds of entities is not by itself sufficient for their existing in different ways. In that case, different kinds of natural substances, e.g. horses and dolphins, or, more radically, horses and gladioli, would also manifest different ways of existing. The thought is, rather, that horses and angels are different orders of beings, not merely different kinds. Because they are different orders of beings, natural substances and angels exist in different ways. Frede does not defend the claim that different orders of beings exist in different ways; nor even does he explicate the thesis that different entities can exist in different ways. He does, however, provide some indication of how these claims might be understood. He claims:

> The substantial forms of sensible substances, in order to be at all, have to be realized in a composite substance that has various non-substantial characteristics, size, weight, shape, colour, etc. Separate substances, on the other hand, exist without matter and without accidents. The unmoved mover, e.g., and quite generally divine substance, are such separate forms.[53]

There are two ideas here. First is the thought that natural substances and angels must satisfy different sorts of criteria in order to exist. Natural substances must be realized in matter if they are to exist; consequently, they must receive the sorts of accidental predicates attendant to matter. Second, then, natural substances exist in such a way as to receive some accidental predications. Angels, by contrast, need not be realized in matter. They therefore lack these sorts of accidents.

Neither of these differences points to any discernible difference in ways of existence. Let us grant that angels exist necessarily, with all and only necessary properties, and that natural substances exist contingently, with contingent properties deriving from their material nature. Let us also grant that no natural substance exists without being realized in matter. So far, these differences are fully compatible with their existing in precisely the same way. That some things which exist are material and others are not, or that some things which exist do so necessarily while others exist only contingently, does nothing to show that there is more than one way

---

[52] In allowing this condition, I do not take a stand on the question of whether definitions of man necessarily make reference to matter. Aristotle takes up this issue *inter alia* in *Metaphysics* vii. 10–11. See Whiting 1986 and Heinaman 1979. The question of whether man is essentially material is in part the question of whether man is a natural substance in the appropriate sense.

[53] Frede 1987, 87–8.

of existing. Put another way, none of these considerations shows that
'exists' signifies distinct properties as it occurs in 'An angel exists' and 'A
horse exists'. Hence, given the features of Aristotle's semantic theory
established earlier,[54] none of these considerations shows that 'exist' has
several senses. Therefore, nothing in Frede's approach establishes the multiplicity all homonyms as such must exhibit.

If this is correct, Frede's treatment of the problem of the unity of
Aristotle's metaphysics exploits the core-dependent homonymy of being
to no good end. He cannot reconcile the universal features of a general
science of being qua being with Aristotle's more narrowly focused discussions of the nature of substance, *ousia*, by treating all beings, and indeed
all substances, as existing in a way explanatorily posterior to the way in
which the divine substance exists. Therefore, we arrive at no clearer understanding of the problem of the unity of Aristotle's metaphysics by appealing to the core-dependent homonymy of being. For we have not moved
beyond the conceptually prior problem of understanding the multiplicity
of being required to establish its core-dependent homonymy.

Our interest in the general problem of the unity of Aristotelian metaphysics has been limited to the role the core-dependent homonymy of
being putatively plays in its resolution. In rejecting Frede's solution, we
have not ruled out other possible solutions; nor even have we ruled out a
solution given in terms of the core-dependent homonymy of being. For
we have not yet seen any reason for denying that being is, in the end, a
core-dependent homonym. Rather, we have seen that one attempt to
establish the homonymy of being fails, because it does not establish its
non-univocity. Consequently, questions pertaining to the priority of some
senses of being relative to others are at this stage idle.

9. 7 THE HOMONYMY OF BEING AND FUNCTIONAL DETERMINATION

Because they are both distinct and associated, core-dependent homonyms
should evince unity in multiplicity. I have argued thus far that Aristotle
has difficulty establishing the multiplicity among beings. Perhaps this is
because I have not attended to another device Aristotle employs for differentiating homonyms—namely, FD, the functional determination thesis. According to FD, an individual will belong to a kind or class *F* if and
only if it can perform the function of that kind or class. Hence, according
to FD, it is both necessary and sufficient for *a*'s being a member of kind *F*

---

[54] See 3. 2.

that *a* have the functional capacity determinative of *F*s. If beings display the functional variations of e.g. bodies, then they too will be homonyms; if they also show the connectedness of e.g. medical things, beings will be core-dependent homonyms.

It is difficult to differentiate beings along the lines FD dictates. Indeed, if FD is the only mechanism available for establishing homonymy, we may be surprised to find Aristotle claiming that beings are homonymous at all. For no function attaches to beings as such. Let us grant for the moment that everything which exists is actual; and let us grant further that everything which is actual is in some sense capable of acting. Yet we cannot infer on this basis that there is some function attaching to beings in so far as they are beings. If not, then the homonymy of being cannot be established in the way Aristotle typically establishes homonymy.[55]

There are two sorts of response Aristotle might adopt. First, noting that FD requires that some entities form a kind only if they have the same function, he may hold that precisely because no function attaches to all things which exist, in so far as they exist, there is no one class comprising them. That is, Aristotle may want to concede that no function attaches to being as such but to use this very fact as a premise to establish homonymy. Hence, he may infer, 'exist' is homonymous because there is no one class of entities, functionally determined, whose binding universal 'exists' signifies.

This is not advisable for him. The doctrine of core-dependent homonymy, as explicated, is a doctrine about metaphysical and semantic multiplicity: the problem is not merely that there is no one universal signified by e.g. 'health', but rather that there are many related universals answering to this term in distinct applications and that they are all definitionally connected to the core notion. In the case of being, we do not seem to have function at all and so trivially do not have a mechanism for sorting beings into kinds. Hence, FD offers no immediate framework for explicating the homonymy of being.

---

[55] So Owen (1960, 198) on one of Aristotle's stock examples for displaying homonymy: 'The example preoccupied him: over and over again in his writings he cites the case of a predicate which is applied both to an original and to a picture or statute; but always—even in works which elsewhere make good use of focal meaning—he cites it simply to illustrate homonymy. His reason for doing so is clear and unvarying. An eye or a doctor, a hand or a flute, is defined by what it does; but an eye or a doctor in a painting cannot see or heal, a stone hand or flute cannot grasp or play. So when they are used in the latter way, "eye" and the other nouns must be used homonymously.' See *PA* 640^{b}29–641^{a}6; *An.* 412^{b}20–2; *GA* 726^{b}22–4; *Pol.* 1253^{a}20–5. Here Owen treats homonymy and 'focal meaning' as mutually exclusive. We have seen that although discrete and associated homonymy are exclusive of one another, they are nevertheless both forms of homonymy. None the less, establishing that core-dependent homonyms are homonyms at all involves establishing first of all that they are not synonyms, and so that they exhibit multiplicity.

Someone might counter that beings as such do have functions. One way of making sense of this notion would be to revert to a semantic conception of the categories, according to which the categories categorize beings of markedly distinct *semantic* types. These types will be, as per FD, individuated by their functions, where functions are restricted to the sorts of semantic roles entities in the distinct categories can play in compositional semantics. Perhaps such functions could permit Aristotle to parcel beings into functionally determined semantic kinds: substances can receive predicates but cannot be predicated; qualities can be predicated as well as receive predicates; quantities provide number; relations relate, etc. This will allow Aristotle to claim that 'exists' is homonymous because it signifies distinct properties in the distinct categories, from a semantic point of view: to say that substances exist will be to say, in part, that there are entities which (in the idiom of Aristotle's early ontology) are neither said-of nor in, that is, which can receive predicates but cannot be predicated; to say that qualities exist is to say that there are entities (again in Aristotle's earlier idiom) that are said-of and in, or both said-of and in, that is, that can function as predicates.

This proposal tries to ground the homonymy of being, in part, in Aristotle's semantic theory. Still, it does not directly try to establish the homonymy of being by adding to the doctrine of categories the premise that being is not a genus.[56] The current proposal is more radical, because it does not merely employ theses of Aristotle's semantic theory in an effort to establish the homonymy of being. Rather, it treats the homonymy of being as itself a semantic thesis, as a thesis describing the markedly distinct semantic functions displayed by sub-propositional components. The general structure of the argument would be as follows:

(1) Substances, qualities, quantities, relations, and so forth[57] exist.

(2) Substances, qualities, quantities, and relations play discernibly distinct functional roles in compositional semantics.

(3) FD.

(4) Hence, substances, qualities, quantities, and relations are distinct kinds.

[56] Someone might seek to establish that being is a genus by appealing to FD. This would represent yet another strategy, though one which would fall prey to the sorts of objections mounted in 9. 8.

[57] The number of categories does not much matter for the present argument, so long as there are more than one. I will proceed with the simplifying practice of mentioning just four in this discussion, since this will suffice for explicating and criticizing the argument.

(5) Hence, 'exists' signifies entities with discernibly distinct semantic functions in 'Substances exist', 'Qualities exist', 'Quantities exist', and 'Relations exist'.

(6) A term is homonymous if it signifies distinct properties.

(7) Hence, 'exists' is homonymous.

This interpretation of the homonymy of being is ineliminably and at root semantic in character. On this approach, 'exists' exhibits precisely the semantic multiplicity all homonymous terms exhibit; indeed, as a defining characteristic of homonymy, we could hardly have expected anything else.

Yet this is just the problem: in grounding the homonymy of being in an unapologetically semantic framework as articulated by FD, we have sacrificed the metaphysical character of the doctrine. More important, even the semantic doctrine is opaque, since it is unclear how (5) can be supposed to follow from (1)–(4). The idea seems to be that 'exists' as it occurs in the various categories admits of paraphrases which reveal an unnoticed homonymy. So, for example,

(a) Substances exist.

yields the paraphrase

(a') There exist things capable of receiving predicates.

whereas

(b) Qualities exist.

yields

(b') There exist things capable of being predicated.

This may be so. Yet it is unclear why 'exists' should be thought homonymous on this basis, when it evidently appears in precisely the same sense in (a), (a'), (b), and (b').

Even if FD parcels members of the categories into distinct semantic kinds, it will not follow that 'exist' as it occurs in the various categories is homonymous. Even if it had, there would be a second insurmountable problem for the present proposal. For being is held to be not merely a homonym but a core-dependent homonym. Even if it could be distinguished along the lines FD commends, the various ways of being would need to be related back of some one source. More precisely, beings, as semantic kinds, would need to satisfy the quite reasonable formal constraint on any account of core-dependent homonymy given in terms of CDH_4.

Yet now, whatever our scruples about the ontic dependency relations between substances and non-substances, we are no longer compelled to mention the category of *ousia* in our account of the semantically individuated categories of being. Relations can relate *relata* outside the category of substance, including relations themselves; quantities can be quantities of non-substances; and predicates generally can be predicated of anything at all. Hence, accounts can be given of these categories without appealing to *ousiai*—even if it is true (as we have in any case nowhere established) that if there were no substances nothing else would exist. Consequently, since Aristotle himself sets the formal constraint, if we understand the homonymy of being principally as a semantic doctrine, then Aristotle's position fails in its own terms. Hence, this approach to the homonymy of being fails along with the others.

9. 8 FROM THE CATEGORIES TO HOMONYMY?

Since an attempt to ground the homonymy of being in some functionally individuated set of categories fails, we may want to reflect on a comparatively general categorial approach. This, at any rate, is what Aristotle would have us do, since he sometimes infers from the doctrine of the categories to the homonymy of being (*An. Pr.* 48^{b}2–4, 49^{a}6–10; *Met.* 1017^{a}22–30). When distinguishing what exists *kath' hauta* (in themselves) from what exists merely *kata sumbebêkota* (coincidentally) in *Metaphysics* v. 7, for example, Aristotle claims:

> Those things are said to be *kath' hauta* that the types of the categories signify; for *to einai* signifies in just as many ways as these types. Since some of the categories signify what the subject is, others its quality, others its quantity, others relation, others activity or passivity, others its place, others its time, 'being' signifies the same as each of these. (1017^{a}22–7)

If being signifies differently in each of the different categories, then the 'is' in 'Socrates is' should signify something other than the 'is' in 'Blueness is'. If we accept the claim that signification is a semantic relation for Aristotle,[58] then the claim translates into the claim that the occurences of 'is' in these statements have discernibly distinct semantic functions. The point can also be put more neutrally, without appealing to Aristotle's semantic theory: what it is for Socrates to be is not the same as what it is for a colour to be. So being for a colour is not the same thing as being for Socrates. In either case, there is a distinction to be marked by 'is' as it occurs in these distinct contexts. If the doctrine of categories reveals these

---

[58] See 3. 3–5.

different senses of 'is', or these different types of existence, or these different ways of being, then the homonymy of being follows directly from a commitment to Aristotle's framework of categories.

I do not see any clear or compelling argument for such a direct inference from the categories to the homonymy of being.[59] Still, there is a sort of philosophical datum here, and Aristotle is right to draw our attention to it. When asked 'What is it?' of a motley group of entities (where 'entities' is very broadly construed), e.g. of Socrates, of being in the Lyceum, and of being four things, one is strongly inclined to say first that these are different sort of things, and second that these different sorts of things are related to one another by some form of priority. That is, one is inclined to say that Socrates is a substance, while being in the Lyceum is not a substance, but rather something we might predicate of a substance. We are, moreover, hardly inclined to say of being in the Lyceum that it is Socrates. What accounts for these inclinations and disinclinations? One conjecture, perhaps a bold conjecture, is that these different kinds of beings *are different ways of being* or that the different kinds of entities *have different ways of being*. That is why we wish to say that some things are substances but not relations, and other things are relations but not substances.

If we focus on this conjecture as an explanation of the indisputable datum that we sort things in roughly the way Aristotle suggests in the *Categories*, we may have the makings of an argument for the homonymy of being. We may perhaps be able to construct an argument for Aristotle by using the resources of the doctrine of categories together with several rudimentary theses of his semantic theory. This is what Grice has attempted to do.[60] He offers us the following reconstruction of an Aristotelian argument for the homonymy of being:

(1) Every simple declarative sentence contains a verb phrase signifying something of something else—i.e. attributing a universal to some subject.

---

[59] Ross (1924, i. 256) seems to accept this inference directly.

[60] Grice 1988. It is worth noting how much disagreement there is among Aristotle's readers on the logical relations between (i) the homonymy of being; (ii) the thesis that there is no genus of being; and (iii) the doctrine of the categories. As we see in the text, Grice wishes to infer (i) from (iii), by relying on (ii). Contrast Barnes (1995, 72): 'Aristotle's argument in Book Beta [of the *Metaphysics*, that being is not a genus] may seem less than compelling; but he also has another reason for denying that entities form a kind or that (in the traditional phrase) "being is a genus". The reason turns on the thought that "things are said to be in various ways"—on the thought that the verb "be" (or "exist") is homonymous.' Barnes wishes to infer (ii) from (i), without reference to (iii). As I shall suggest, the lack of consensus about the logical priority of (i) and (ii) in Aristotle's thought stems from the fact that Aristotle himself is unsettled on just this question.

(2) Proposition (1) holds true for existentials (as instances of simple declarative statements).

(3) Hence, existentials attribute universals to subjects.

(4) If 'exist' signified a single universal, it would signify a generic universal (since the different categories would be different ways of being, and so different species of being).[61]

(5) Existence is not a genus.

(6) Hence, 'exist' does not signify a single universal.

(7) Hence, 'exist' signifies a plurality of universals.

Grice adds a few steps to show that the plurality of universals signified correspond precisely to the ten categories of being:

(8) If 'exist' signifies a plurality of universals, that plurality should satisfy two conditions: (i) it should be as small a plurality as possible; and (ii) each universal in the plurality should attach essentially to whatever it primarily attaches to.

(9) The only set of universals satisfying both conditions are the category-heads themselves.

(10) Hence, 'exist' signifies precisely the plurality of universals answering to the ten categories (that is, being a substance, being a quality, etc.).

Many philosophers will worry about the second premise, holding that existence is so semantically discontinuous in its functions with standard predicates that it would be a grievous confusion to think of it as a predicate at all. This is not, however, my worry.[62] Rather, I am concerned about premise (4), and on two grounds. First, it is not the case that all ways of *F*-ing are species of some genus *F*. Some waiters wait courteously and some rudely; some wait efficiently and some inefficiently; some wait happily and others morosely. I do not see why we should think there is a genus 'waiter' with a species answering to each of these ways of waiting. Second, even if we allow a species $F_1 \ldots F_n$ for every way of *F*-ing, I do not think we can, in the context, use this as a reason for believing there are a multiplicity of universals corresponding to the general (genus) term '*F*'. Let us admit gimpy walkers, speedy walkers, and strolling walkers as species of the genus 'walker'. Should we now say that 'walk' is homonymous and

[61] Grice (1988, 178) says: 'This step has been supplied by me.'
[62] Aristotle is willing to predicate being; see e.g. *De Interpretatione* 11.

signifies a plurality of universals? Certainly further argumentation is required here. If we saw a man and a woman walking down the road together, and if the woman were limping ever so slightly, nothing would tempt us to say 'He is walking*, whereas she is walking**'. For these reasons, I do not think we can rely on the premise Grice supplies as the bridge from the categories to the homonymy of being.

Indeed, I am inclined to think Grice is wrong to assume that we could structure any non-question-begging argument directly from the doctrine of the categories to the homonymy of being. What is at issue is whether we can detect distinct senses of 'exist' by surveying the categories of being: Platonists will deny that we can, insisting, as seems reasonable, that all the categories of being are categories of things that exist, no matter what the ontic dependency relations between them may be; and what it is for them to exist in each case comes to the same. Grice's Aristotle has countered only by observing that since the items in distinct categories exist only by being items in those categories, we should refrain from thinking of them as existing *in the same way* and so should refrain from thinking of them as instantiating the single generic universal. Thus far, we have at best a stalemate, and an argument unlikely to appeal to anyone not already converted.

Moreover, despite its initial appeal, Grice's construal does not capture the connection Aristotle sees between the categories and the homonymy of being. Aristotle himself provides a much more technical argument in *Metaphysics* iii. 3, a passage where he sketches an *aporia* for those who think that the most basic principles (*archai*) of things are genera. If the most basic principles are genera, then, Aristotle suggests, these may be either the highest forms of categorization of things, the *summa genera*, or the lowest, the *infimae species*. In tracing out the consequences of the first possibility, he claims:

If the universal is always more of a principle, then it is clear that the highest among the genera <will be principles>. For these are said of all things (*legetai kata pantôn*). There will be, consequently, as many principles of beings as there are primary genera, so that both being and one (*to te on kai to hen*) will be principles and substances, for these, most of all, are said of all beings. But neither one nor being can be a single genus of beings.[63] For it is necessary that the differentiae of each genus be and that they each be one; yet it is impossible either for the species of the genus to be predicated of their own differentiae or for the genus to

---

[63] Reading *ouch hoion te de tôn ontôn hen einai genos oute to hen oute to on*, with the dominant manuscript tradition. The alternative, *oute to hen oute to on einai genos* in A[b] states more simply that being cannot be a genus. Aristotle's point here need only be the more restricted one that there is no *one* genus of being, even if there may be several related genera.

be predicated <of its own differentiae> in the absence of its species. Hence, if either one or being is a genus, no differentia will either be or be one. However, unless they are genera, they will not be principles, if indeed the genera are principles. (*Met.* 998ᵇ17–28)

The basic structure of Aristotle's argument is clear. In setting the first half of the *aporia*, Aristotle suggests that if we adopt the thesis that the basic principles (*archai*) of things are the most general universals, we end up endorsing something impossible, namely that being and one are genera of things. A brief argument should demonstrate why this is impossible and so, ultimately, why the *archai* of things cannot be the highest genera.

Although advanced in an aporetic passage, this argument clearly holds some appeal for Aristotle.[64] Indeed, in showing why no one should hold the thesis that the *archai* of things are the *summa genera*, Aristotle appeals to his own conviction that being is not a genus. He does not convincingly demonstrate that anyone who accepts the thesis that the basic principles are the *summa genera* must also hold that these genera are being and one.[65] He none the less deploys a technical argument with the conclusion that being cannot be the highest genus; this argument is free-standing, and if it establishes that being cannot be the highest genus, it also shows that it cannot be any one genus at all.[66]

Aristotle's argument that being cannot be any one genus appeals to technical features of his taxonomical method quite distinct from the more general features of the doctrine of categories to which Grice appeals.[67] The argument may be best displayed in the form of two related *reductiones*:

---

[64] See Ross 1924, i. 235.

[65] Someone who holds that the *archai* are the *summa genera* could presumably agree with Aristotle that being and one are not *genera*, only to insist that the *summa genera*, whatever they turn out to be, are the *archai* sought. In this sense, Aristotle's argument fails rather early on. Yet it does not therefore lose the interest it holds for the present discussion. For once the illicit claim is made, it becomes possible to determine independently whether being and one are genera, and this is what Aristotle proceeds to do.

[66] This assumes on Aristotle's behalf that being could not be a single genus ordered under some higher genus. Not everyone will grant this; perhaps the Stoics mean to deny it with a distinction between *einai* and *hupistasthai*. (This at any rate is how Alexander of Aphrodisias represents them in a polemical passage, *In Ar. Top.* 301, 19–25 (= *SVF* 2.239).) This distinction is not in Aristotle, and his point in the present passage seems insensitive to it as a possibility.

[67] Grice at least provides *an* argument. Some others seem to hold that one can take the claim that being is not a genus for granted and infer directly to the homonymy of being. So Owen 1965a, 265: 'First, no category is a species of any other: substances are not a kind of quality nor qualities a kind of substance (e.g. *Met.* 1024ᵇ15, 1070ᵇ3–4). Second, no category is a species of *being* or *what there is*, for there is no such genus as *being* (e.g. 998ᵇ22–7). So it seems that the verb "to be" in its existential rôle enjoys a number of irreducibly different senses.'

(1) Suppose being and one are genera.

(2) Every differentia of a genus (a) exists[68] and (b) is one.

(3) Hence, (a) the differentiae of being will (i) exist and (ii) be one; and (b) the differentiae of one will (i) exist and (ii) be one.

(4) If (3a. i), then species under the genus being will be predicated of their own differentiae.

(5) If (3b. ii), then species under the genus one will be predicated of their own differentiae.

(6) No species can be predicated of its own differentiae.

(7) Therefore, neither (3) nor (4) is true.

(8) Hence, either (1) or (2) is false.

(9) Proposition (2) is true.

(10) Hence, (1), our original supposition, is false.

A second impossible consequence yields the same result. The second *reductio* diverges from the first at the fourth step:

(1) Suppose being and one are genera.

(2) Every differentia of a genus (a) exists and (b) is one.

(3) Hence, (a) the differentiae of being will (i) exist and (ii) be one; and (b) the differentiae of one will (i) exist and (ii) be one.

(4′) If (3a. i), the genus 'being' will be predicated of its differentiae in the absence of its species.

(5′) If (3b. ii), the genus 'one' will be predicated of its differentiae in the absence of its species.

(6′) It is not possible for a genus to be predicated of its own differentiae in the absence of its species.

(7) Therefore, neither (3′) nor (4′) is true.

(8) Hence, either (1) or (2) is false.

(9) Proposition (2) is true.

---

[68] Resolving Aristotle's *einai* into 'exist' here represents a decision about the point of the argument not yet reflected in the translation given above. This is in keeping with our attempt to find a strong interpretation of the homonymy of being, such that the existential sense of 'is' is itself a multivocal. See 9. 1.

(10) Hence, (1), our original supposition, is false.

These two arguments equally situate the putative impossibility of there being a single genus of being (or one) in the context of the conditions Aristotle believes obtain in instances of intra-categorial predication. In each case, the supposition that being (or one) is a genus ultimately violates some such condition.

Neither argument makes entirely clear the precise condition being violated. I will begin with the second argument, since the first introduces complications additional to those in the second. Step (1) is the supposition. Step (2) merely claims that for any given differentia, it will be one thing and will exist. Thus, if 'two-footed'[69] is a differentia in the genus 'animal', there is something, two-footedness,[70] which is one thing. Neither of the assumptions is uncontroversial,[71] and each requires independent consideration. They may be granted in the present context, however, since a sufficiently relaxed version of each can be defended, and this is all that is needed for the argument to proceed. Step (3) is merely a specification of (2). Proposition (2) tells us that for any genus, the differentiae falling under it have the properties of existing and of being one.[72] Proposition (3) claims

---

[69] I will adopt the convention of placing species, genera, and differentiae in quotation marks wherever this will allay possible confusion. In some cases this is not an issue.

[70] The status of the differentia is difficult and disputed. Aristotle suggests that in assessing an argument, it is always necessary 'to see if the differentia offered signifies some this, for every differentia seems to indicate a quality' (*horan de kai ei mê poion ti tode sêmainei hê apodotheisa diaphora; dokei gar poion ti pasa diaphora dêloun; Top.* 144ᵃ20–2, omitting the *alla* in line 20 with C). The language here is somewhat guarded. Perhaps more important, Aristotle uses the principle that the differentia is a quality in an argument refuting those who claim that being aquatic and being terrestrial cannot be differentiae, since fish end up on land now and again, people swim, and so forth (*Top.* 144ᵇ32–145ᵃ2). Aristotle responds that 'aquatic does not signify where <something is> or in what <it is>, but rather <signifies> some quality' (*ou gar en tini oude pou sêmainei to enudron, alla poion ti;* 144ᵇ35–6; cf. 122ᵇ12–17, 128ᵃ20–9, 139ᵃ28–31). Still, if it is a quality, the differentia is unique. First, Aristotle regularly (though not invariably) matches a differentia term in gender with that of an unstated genus, thus making predications of the differentia truncated predications of the genus, delimited in the appropriate way. Thus, 'Man is two-footed' compresses 'Man is *a* two-footed animal', where the predication is nominal rather than adjectival, as we might expect for predications of qualities. Irwin (1988, 64–6, 508 nn. 46 and 47) objects to the view that the differentia is a quality in the sense required in the *Categories* 3ᵃ21–8. Qualities inhere; differentiae do not. These objections do not preclude our treating the differentia as a quality in the present context, since the arguments do not require that differentiae inhere in the technical sense of the *Categories*. See also Ackrill 1963, 86; Granger 1984; Trendelenburg 1846, 56; Bonitz 1848–9, 258; Leszl 1970, 57–9.

[71] The first claim requires defence, since it insists on the differentia having existential import, thus ruling out the possibility of there being kinds of nonexistent things (say, kinds of monsters). The second claim is less controversial, so long as the composite properties are treated as individuals.

[72] I will adopt the convention of talking about existence as a property in this discussion, but only as an expository convenience. The argument could be more cumbersomely stated in other terms.

that if being is a genus, its differentiae must have these properties, so that the differentiae of this genus will both exist and be one thing. The same holds true of the postulated genus 'one'.

The real work of the argument begins at (4'), with the curious locution that on the assumption of (1), the genus could be predicated of its differentiae 'in the absence of its species' or, simply, 'without its species' (*aneu tôn autou eidôn*; *Met.* 998ᵇ25–6). Aristotle's point is initially somewhat opaque. Some illumination comes from the *Topics*, where he is concerned to specify in a self-conscious way the structure of adequate definition. Of particular help is *Topics* vi. 6, a chapter in which Aristotle elucidates the requirements for intra-categorial predication, and clarifies the way predications involving the differentia function in essence-specifying definitions.[73]

There Aristotle argues that the genus cannot be predicated of the differentia (*Top.* 144ᵃ31–ᵇ3), although he does not mention in addition that the problem stems from predicating the genus of the differentia without including the species. The point, however, seems to come to the same:[74] although one could predicate animal of being a two-footed man, one could not predicate it merely of being two-footed. Presumably, the addition of the species makes the predication true, since 'animal' states the essence of man, while the differentia is superfluous, since it states, at best, a necessary but non-essential property of man,[75] what Aristotle elsewhere calls an *idion*.[76] If one predicates the genus in the absence of the species, however, difficulties emerge. Then an illicit predication results, in the form of 'being two-footed is an animal'.

---

[73] After canvassing general forms of definition in *Topics* vi. 1–3, *Topics* vi. 4 turns the discussion towards the question of 'whether or not one has stated and defined the essence' (*Top.* 141ᵃ24–5). *Topics* vi. 6 presumably falls under the same rubric. This is the fifth of the five parts of the discussion Aristotle outlines at the beginning of *Topics* vi. 1.

[74] Ross (1924, i. 235) implicitly takes the argument this way, since he thinks that Aristotle's point is that 'the genus cannot be predicated of its differentiae'.

[75] Here again the status of the differentia is difficult. Aristotle does not think that his standard example of a differentia, being two-footed, is a necessary property of human beings (*Top.* 134ᵇ5–7). Hence, Irwin (1988, 508 n. 47) reasonably concludes that 'it is not a proprium, since a proprium is a necessary property and it is possible for man not to be a biped'. Still, Aristotle elsewhere claims that 'no differentia is among the things belonging co-incidentally (or, accidentally), just as no genus is; for it is not possible for the differentia to belong and not belong to something [at different times]' (*oudemia gar diaphora tôn kata sumbebêkos huparchontôn esti, kathaper oude to genos; ou gar endechetai tên diaphoran huparchein tini kai mê huparchein*; *Top.* 144ᵃ24–7). Perhaps the inference to draw is that being two-footed is not a differentia in the strictest sense, and is used principally as an illustration.

[76] *Top.* vi. 3 and 4; cf. 102ᵃ18–24. On characteristics of *idia* see Irwin 1988, 507 nn. 38 and 39; Copi 1954; Moravcsik 1967; and Joseph 1906.

Aristotle lodges two complaints here,[77] the first of which is itself rather obscure. Aristotle says, 'If animal [the genus] is to be predicated of each of its differentiae, many animals will be predicated of the species' (*ei gar kath' hekastês tôn diaphorôn to zô(i)on katêgorêthêsetai, polla zô(i)a tou eidous an katêgoroito*; *Top.* 144ª36–7). Second, since every animal is either an individual (*atomon*)[78] or a species, then if the differentiae are animals (*eiper zô(i)a*), they too must be either individuals or species (*Top.* 144ᵇ1–3). This second argument is relatively straightforward.[79] The differentia is a quality, not an individual or a species. A genus is a genus only of species and their individual members. This is why both 'Socrates is an animal' and 'Man is an animal' count as acceptable predications. Hence, we force the differentia into the wrong category by trying to predicate the genus of it in the absence of the species whose differentia it is.

The first argument is not so brief, and not so easily unpacked, but holds more promise for help in understanding our initial problem, namely, why we cannot postulate the genera 'one' or 'being' of their differentiae. If the genus 'animal' is predicated of its differentiae, then 'many animals' (*polla zô(i)a*) will be predicated of the species. Expanding Aristotle's compressed suggestion, the argument seems to appeal implicitly to homonymy.[80] If being two-footed (or being terrestrial or being aquatic) counts as being an animal, that is, if 'Being two-footed is an animal' has a true reading, the notion of 'animal' employed must be significantly unlike the notion employed in 'Callias is an animal'. Hence, there will be a multiplicity of senses or properties potentially predicated of Callias. This is just to say, however, that animal is homonymous and not synonymous. For the account of being (*logos tês ousias*) of animal will not be the same in the projected predications.

Aristotle's point here should probably be understood as a formal or counterfactual point. He is not claiming that 'Being two-footed is an animal' employs a clear sense of 'animal' other than the sense attaching to 'animal' in standard applications. On the contrary, his point is precisely that such predications are illicit. If one *were* to predicate 'animal' of its various differentiae, then one would render 'animal' homonymous by inventing non-standard senses of the term. For in order even to make the

---

[77] Waitz (1844–6, ii. 500) also holds that Aristotle's arguments are distinct: 'Aliud argumentum affertur, quo probetur genus non praedicari de differentia.'

[78] Although this is not Aristotle's standard locution for individuals, he does use it on some other occasions: *Met.* 995ᵇ29, 998ᵇ16, 1018ᵇ6, 1034ª8, 1058ª18, 1059ᵇ36.

[79] Waitz (1844–6, ii. 500) seeks a parallel in *Topics* 144ª8; but the discussion there concerns not whether the genus can be predicated of the differentia, but whether someone has mistakenly taken the genus to *be* the differentia.

[80] Waitz (1844–6, ii. 500) plausibly suggests that *polla zô(i)a* should be understood as *pollakis to zô(i)on*.

predications truth evaluable, one would need to postulate novel senses for each case. The procedure would result in some indefinite number of senses of 'animal', and in any given predication, it would always be in principle an open question which of these senses was to be understood. In this way, one would be predicating 'many animals' of the species. Hence, predicating the genus of the differentia renders the genus term homonymous.[81]

If this is correct, Aristotle envisages two distinct ways of supporting (6'), the crucial premise of the second *reductio* argument seeking to establish that being cannot be a genus. Proposition (6') holds that it is not possible for a genus to be predicated of its own differentiae in the absence of its species. I have suggested that this is best understood as collapsing into the claim that the genus cannot be predicated of the differentia, a view Aristotle defends with two distinct arguments in *Topics* vi. 6. Hence, the plausibility of (6'), and so of the second *reductio*, rests entirely on the defensibility of these claims.

The comparatively straightforward argument is not compelling. Aristotle argues that if the genus is predicated of the differentia, the differentia will itself be either an individual or a species (*Top.* 144^{b}1–3). If the genus is to be predicated, however, it must itself be a certain sort of property.[82] But since it is the kind of property that specifies the essence, the genus will need to be a privileged property of a certain sort. The genus, then, is a *privileged$_e$* property, where all privileged$_e$ properties are such that if predicated at all, they are predicated essentially of substances. So, all and only substance sortals are privileged$_e$ properties. Consequently, if something is a privileged$_e$ property, anything instantiating it must be a substance. All substances are either individuals or species.[83] Hence,

---

[81] This is something Aristotle will want to deny: 'It belongs to substances and differentiae that all things <named after them> are spoken of synonymously' (*Cat.* 3^{a}33–4).

[82] I use the term 'property' here in a neutral sense, ignoring Aristotle's preference for treating the genus as a nominal rather than adjectival predication.

[83] Aristotle holds that the species is a secondary substance in the *Categories*. 'Substance—most strictly and primarily, and most of all—is neither said of any subject nor in any subject, e.g. some man or some horse; the species, in which what are called substances primarily belong, are called secondary substances, as are the genera of these species. For example, some man belongs in the species man, while the genus of this species is animal, so these, that is, man and animal, are called secondary substances' (*Cat.* 2^{a}11–19). In the *Metaphysics*, by contrast, he claims that 'it is clear that none of the things belonging universally is a substance, and none of the things predicated in common signifies a this, but a such' (*phaneron hoti ouden tôn katholou huparchontôn ousia esti, kai hoti ouden sêmainei tôn koinê(i) katêgoroumenôn tode ti, alla toionde*; *Met.* 1038^{b}35–1039^{a}2). Hence, unless he implicitly means that none of the things predicated universally is a *primary* substance, Aristotle has changed his mind about substance. The matter is disputed. If he has changed his mind, as I believe, then the claim in the text needs to be qualified in the appropriate ways. If he has not, it can stand as it is. For arguments that the substantial form of *Metaphysics* vii is to be identified with the species of the *Categories*, see Ackrill 1963, Owen 1965a,

Aristotle may conclude, if it receives the genus as a predicate, the differentia must be an individual or species.

If we focus on Aristotle's preferred categories, this second argument seems plausible. We would not want to regard the differentiae 'being two-footed' or 'being aquatic' as species of animals or as individual animals. We think that the genus 'animal' counts as a privileged$_e$ property. It does not follow, however, that Aristotle can extend this point directly to a postulated genus 'being' or 'one'. If there is a *summum genus* of being, then any given differentia falling under it will exist and so will have its genus predicated of it. Yet nothing requires that this differentia be treated as a species or individual other than the general principle that if *G* is a genus predicated of *S*, *S* must be either a species or an individual. Put another way, Aristotle's argument is compelling only on the assumption that every genus is a privileged$_e$ property. And of course this principle would be false if being were a genus, since being would not be a privileged$_e$ property. In seeking to apply this principle to a postulated category of being by insisting that being would be a privileged$_e$ property, Aristotle presumes that we have independent reason for supposing that it would hold true even in the case of being. Certainly the brief argument under consideration provides no such independent grounding. Hence, Aristotle's second argument for (6′) makes no progress.

The first argument initially holds more promise than the second, since it does not presume that what holds for Aristotle's preferred genera must also hold for a postulated genus of being. Instead, Aristotle wonders how we must regard the genus if it is to be predicated of both the species and the differentia. If someone insists that 'Being two-footed is an animal' is a legitimate, well-formed predication, then, Aristotle insists, this must be because 'animal' is homonymous. For on the assumption that the genus is predicated of a non-substantial differentia, the predication 'animal' is no longer a privileged$_e$ predicate. Hence, it must be some other, presumably related predicate, to be further specified by the one employing it in the new, non-standard way. So if the genus 'animal' is predicated of its differentiae, 'many animals' (*polla zô(i)a*) will be predicated, and 'animal' will be spoken of in many ways (*pollachôs legomenon*).

Considered in itself, this line of reasoning is at least initially attractive.

Woods 1967, and Furth 1988. For arguments that the substantial form of the *Metaphysics* is either not a universal but nevertheless predicated in common or else a universal restricted in its application to matter, see Code 1984 and 1986, Driscoll 1981, and Lewis 1985 and 1991. For arguments that substantial forms are particulars or individuals, see Sellars 1957, Irwin 1988, Whiting 1986, Shields 1988a, and Frede and Patzig 1988. For a clear summary of the issues, see Wedin 1991.

Yet it is difficult to appreciate how Aristotle could appeal to just this argument in the current context. He is trying to show that (6′) is true. If (6′) is true, then it is not possible to predicate a genus of its differentia. This, in turn, Aristotle wants to show because he wants to show that being is not a genus, which is, finally, necessary for showing that being is homonymous. If this is correct, the current train of inference is deeply problematic:

(1) Being is not a genus.

(2) If being is not a genus, then (on the condition that certain ancillary conditions are met) being is homonymous.

(3) Hence, being is homonymous.

But the first premise of this argument is defended as follows:

(1) If being were a genus, then it would be possible to predicate the genus 'being' of its differentiae.

(2) If the genus 'being' were predicated of its differentiae, then being would be homonymous.

(3) Being is not homonymous.

(4) Hence, being is not a genus.

Now, one may want to defend either of these arguments, taken individually. But it will not be possible to defend the first premise of the first argument by appeal to the second argument; for the third premise of the second argument is a premise Aristotle rejects; indeed, it is actually a straightforward denial of the conclusion of the first argument.

This puts Aristotle in an awkward and untenable situation. He wishes to show that being cannot be a genus. To do so, he must show that it is illicit—as a general principle—to predicate the genus of the differentia. In arguing for this general principle, he insists that *if* genera were to be predicated of differentiae, it could only be because those genera were predicated homonymously. Extending his point to being, then, involves him in saying, counterfactually, that if being were predicated of the differentiae falling under it, it would be homonymous. In order to block this inference, then, he needs to deny this claim by insisting that being is not homonymous, which is, of course, just a denial of the first argument's conclusion. Hence, either he cannot defend the general principle that genera cannot be predicated of differentiae, or Aristotle must deny the very conclusion he sets out to establish, namely, that being is homonymous. He ends up with a tangled and internally inconsistent set of claims which he cannot jointly maintain. Hence, this attempt to establish the non-univocity of being from the thesis of the categories fails, and in a markedly unhappy way.

Summing up, we see that Aristotle structures an aporetic argument in *Metaphysics* iii. 3 for the claim that genera cannot be the basic principles, or *archai*, of things. He contends that the defender of this position will be forced to select being and one as the most plausible candidates for this status. Since being and one cannot be genera, the proposal fails. Aristotle defends his thesis that being cannot be a genus by means of two distinct *reductio* arguments. Thus far, we have seen that the second of these arguments fails. The claim that being is not a genus is either unsubstantiated by the arguments Aristotle provides, or implicitly appeals to a denial of the homonymy of being itself and so lapses into internal inconsistency. Therefore, any attempt along these lines to establish the homonymy of being by appeal to the doctrine of categories fails.

We have not, however, examined the first *reductio*. Like the second *reductio*, the first appeals to technical features of Aristotle's taxonomical methods rather than the more general features of the doctrine of categories to which Grice adverts. The argument proceeds along precisely the same lines as the second, except that the hypothesis that being is a genus is held to lead to a distinct absurdity. Here the problem is that the assumption that being is a genus leads to the species being predicated of the differentia, something no one can accept.

The argument's first three steps proceed as before: (1) is the assumption; (2) may once again be granted, as may (3), its specification. Steps (4) and (5) present new difficulties. They do not parallel (4′) and (5′) very closely. Proposition (4) holds that if the differentia of a postulated genus 'being' is itself a being, then the species of that genus will be predicated of a differentia. Proposition (5) makes an analogous point about being one. If a differentia of a postulated genus 'one' is itself one, then the species of that genus will be predicated of that differentia. These premises are not as transparent as (4′) and (5′), because in those cases, the postulated genera were immediately predicated of the species. If a differentia in the genus 'being' is a being, then the genus is predicated of it; similarly, if a differentia in the category 'one' is itself one, then its genus is predicated of it. Premises (4) and (5) are not quite so easily explicable, since there is no reason to suppose that the species of the postulated genera will be immediately predicable of the differentiae in just this way.

Still, it might be thought that there is a mediated way in which Aristotle could show that, by there being a genus predicated of the differentia, some species will *ipso facto* also be predicated of that differentia. To begin with a neutral case, the genus 'animal' is predicated of an *x* other than the species ordered under it only if one of those species is also predicated of *x*.[84] So 'animal' is predicated of an individual human being only if the species 'man' is predicated of that human being. Now, if we suppose that there is a genus of being, we think there are species of various sorts ordered under it. These species will be demarcated from one another in the normal way, that is, by their various differentiae. These differentiae will themselves be beings and will each be one, by (2) and (3). Hence, it will be true of them, as we saw in our account of the second *reductio*, that the genera 'being' and 'one' will be predicated of them. (It will be true to say that each is a being, and one.) If this is correct, then we already know that some species must be predicated of them. This will follow, at any rate, if a genus is predicated of something other than a species only if some one of its species is simultaneously predicated of that thing.

If Aristotle's inference is to be supported in this way, then both (4) and (5) are at least explicable. If, by contrast, it is not, then it seems to be altogether inexplicable, and the argument fails at this point. On the assumption, then, that (1) forces us to concede that the species will be predicated of the differentia, we can turn to (6), the claim that it is not possible for a species to be predicated of its differentia.

Although nowhere explicated or defended in *Metaphysics* iii. 3, (6), like (6′), finds support in *Topics* vi. 6. There Aristotle discusses the problem with this sort of intra-categorial predication:

Similarly, one must inquire also whether the species or anything below the species is predicated of the differentia. For this is impossible, since the differentia is said of a wider range <of things> than the species (*epeidê epi pleon hê diaphora tôn legetai*). Further, if it is to be predicated <of the differentia, it will follow that> the differentia is a species, since what is predicated of it is one of the species. For if man is predicated of it, it is clear that the differentia is a man. (*Top.* 144^b4–9)

Once again Aristotle provides two brief ancillary arguments, first that the differentia is said of a greater range of entities than the species, and second that if it receives the species as a predicate, the differentia will itself be a species.

The second argument is clearly unsound as stated, even in Aristotle's own terms. Certainly *a*'s having a species predicated of it is not sufficient for *a*'s being a species itself. The individual members of a species, for example, individual humans, have the species predicated of them even though they are obviously not themselves species. Perhaps Aristotle

---

[84] Aristotle does not offer exactly this judgement. He says, rather, that 'animal' will not be said of man unless it is also said of an individual man (*Cat.* 2ª36–^b1). Still, Aristotle can hardly believe for his preferred species and genera that a genus could be predicated of an individual member of a species without also being predicated of the species, e.g. that 'animal' could be predicated of some man without also being predicated of man.

assumes, correctly, that differentiae are not sub-specific. This would permit him to infer that since the differentia does not have the species predicated of it by being a member of the species, it can only do so by itself being a species. Although this is of course a result we would not want, it is not clear why we would need to embrace it. For Aristotle has not shown that anything not sub-specific which has the species predicated of it is itself a species. Indeed, it is hard to see how this principle could be anything but tailored for the argument in question. It certainly does not follow from any general features of his taxonomical method.

The first argument is similarly inconclusive. It is true that the differentia has a wider extension than the species. If 'aquatic' is a differentia, it will apply to fish of various species. Hence, to say that 'being aquatic' is a trout is to say something false, if it is to say anything at all. If it means that the quality of being aquatic has the property of being a trout, it says something false. If it means that whatever is aquatic is a trout, again it says something false, or perhaps nothing at all. We end up, therefore, with a predication which is either ill-formed or obviously false. If we treat it as false, this will be because 'the differentia is said of a wider range <of things> than the species' (*epeidê epi pleon hê diaphora tôn legetai*; *Top.* 144^{b}6).

Applying this to the projected genera of being and one, we should therefore expect that any predication of the differentia by the species yields either nonsense or an obvious falsehood. But no such results follow.[85] On the contrary, if we move beyond the species and genera of biological kinds, as Aristotle himself certainly does (*Top.* 143^{a}15–19), it is easy to construct examples where we do not have these sorts of consequences. First, if there is a genus 'abstract entity', one species under it will be sets, marked off from other species of abstract entities by the differentia 'being extensional'. The differentia will have the property determined by the genus, namely being an abstract entity, even though we are not compelled to predicate 'being a set' of the differentia 'being extensional'. Still, it may be objected, there must be *some* species predicated of the differentia in virtue of which the genus is predicated of it. Indeed, in this case at least, there is: the species 'property'. But now we are compelled to say only that the differentia 'being extensional' is a property. This is not only meaningful but also true. Focusing more narrowly on the species 'quality' and the differentia which marks it off, namely 'being intensional', we again run into no difficulty. Here it will be true to say that the species is predicated

---

[85] This is something Aristotle evidently appreciates: 'The differentiae of relatives are themselves relatives' (*Top.* 145^{a}13–14).

of the differentia, that is, that the differentia is intensional. But this again is not only meaningful but also true.

Hence, the general principle which is supposed to show that we run into difficulty in the proposed sort of intra-categorial predication is not sufficiently general.[86] In some cases, we are not compelled to predicate the species of the differentia in virtue of the genus's being predicated of the differentia, in which case (4) in our original argument is false. In other cases, we may predicate the species of the differentia with no threat of nonsense, or even falsity, in which case (6) is false. Hence, the general claim that it is impossible that the species be predicated of the differentia fails. Although reasonably assuming that this obtains for some range of cases, Aristotle has not shown that it holds generally. Without showing that it holds generally, Aristotle cannot appeal to it in endeavouring to show that it fails in a special case, namely, the case assumed in (1), where being is a genus—indeed, the *summum genus*.

Aristotle does not, then, establish that being cannot be a genus. Some of his arguments presume that principles applying to the genera he favours extend automatically to a postulated genus of being. Because a genus of being need not respect all principles true of lower-order genera, these arguments fail. They show at most that *if* it is a genus, being is anomalous relative to some other genera, and this conclusion is too weak by itself to establish the homonymy of being. Other arguments do not rely on this illicit form of extension. Yet because they counterfactually deny the homonymy of being itself, these arguments are untenable; they seek to

---

[86] Loux (1973) recognizes some of the problems with this argument but does not appreciate that Aristotle cannot generalize from his preferred categories to postulated genera of being and oneness. Seeing that Aristotle's own taxonomical arguments fail (for reasons other than those I have provided and which I do not endorse), Loux seeks to supplement Aristotle's argument with another, of his own, that the whole notion of a genus is essentially classificatory, and that since the transcendentals, of which he thinks being and goodness are representative (226), cannot classify, they cannot be genera (230–1). Thus, for every genuine genus term, 'there must be competing, incorrect answers' (230), that is, it must be possible to misplace an entity in a genus to which it does not belong (e.g. 'It is not a marsupial, but a true mole'). But everything is one and a being, so these cannot be genera. Some observations: (i) this argument assumes that being and oneness are in fact transcendentals, but it is doubtful that Aristotle treats them this way; (ii) certainly at least 'one' and probably 'being' are classificatory in the sense determined, since there are things which are not one (e.g. someone might have thought that the twins were one, when they are two, not one; as for being, Aristotle mentions a class of non-beings, *ta mê onta*, though in a dialectical context, Alexander 80. 16–18, where one might have mistakenly thought they existed); and (iii) to assert in any case that a genus must be classificatory would be a hollow victory, since we would establish a key premise in the argument for the homonymy of being by fiat. Strikingly, and in any case, Loux argues that even granting the claim that being is not a genus, we are not entitled to infer the non-synonymy of being (231–9). Here, I am sympathetic to much of what he says.

establish the homonymy of being by showing that being cannot be a genus, which in turn they hope to establish ultimately by insisting that being could be predicated of its species only if it were homonymous.

If this is correct, the more technical arguments developed in the context of Aristotle's taxonomical method do not show that being cannot be a genus. Nor, as we have seen, does the argument Grice structures on Aristotle's behalf. One cannot appeal immediately to general features of the categories or the semantic theory they presuppose to show that being is not a genus. On the contrary, in neither approach are we given a compelling reason for supposing that being is not a genus. Without this link, neither strategy ultimately succeeds. Indeed, since these are the only two forms of argument attempting to establish the homonymy of being based upon the doctrine of categories, no such argument succeeds. Therefore, Aristotle cannot establish the homonymy of being by appealing only to the doctrine of categories.

9.9 A GENERAL PROBLEM ABOUT THE HOMONYMY OF BEING

The argument thus far has been an argument by cases. I have considered the principal groundings for the homonymy of being and have concluded that none supports a doctrine which is both recognizably Aristotelian and philosophically defensible. Most of the approaches canvassed suffer from attempting to explicate the core-dependent character of being without first having established its non-univocity. Even if we assume that non-substances depend for their existence on the existence of substances (something, I reiterate, we have in any case nowhere conceded) and infer on this basis that substances are ontologically prior to other sorts of beings, we will not have shown that non-substances exist any less than substances, or in any way other than the way substances themselves exist. In short, ontological priority is not sufficient for the non-univocity of being.

Supposing the individual arguments mounted thus far are cogent, the critical treatment of the homonymy of being offered here is nevertheless inexhaustive in the way of all arguments by cases. It is only as compelling as it is complete, and it cannot be determined a priori that no further approach to the non-univocity of being is available. This is not, however, to concede that the treatment provided fails to challenge the truth of the doctrine. On the contrary, to the extent that the approaches considered represent the dominant interpretations of Aristotle's central application of his doctrine of homonymy, the arguments offered at the very least prescribe reassessment of the defensibility of the doctrine as applied to being.

This conclusion can be strengthened. By reflecting on the connected ways in which the various approaches considered have failed to establish the homonymy of being, one can detect systematic problems which support a more abstract, more comprehensive rejection of the doctrine. The general argument proceeds by showing not only that has Aristotle failed to establish non-univocity in the case of being but that he cannot establish it. Recall that establishing core-dependent homonymy involves a three-stage process: (i) one must demonstrate non-univocity; (ii) one must then establish association; and finally (iii) one must establish core-dependence in line with CDH_4. So far I have argued that the principal arguments for (i) fail. I now argue that there is a reason for this, namely that being is univocal and so not a core-dependent homonym.

The argument for this conclusion is direct.

(1) Two F things are non-synonymously F only if they are incommensurable as Fs.

(2) Beings are always commensurable as beings.

(3) Hence, beings are not non-synonymously Fs.

(4) The distinction between homonymy and synonymy is exhaustive.

(5) Hence, beings are always synonymously beings.

(6) If beings are always synonymously beings, then they are univocally beings.

(7) Therefore, since beings are core-dependent homonyms only if they are non-univocal, beings are not core-dependent homonyms.

The conclusion (7) is simply a way of saying that Aristotle cannot meet the first condition of establishing the core-dependence of being.

Most of the premises of this argument are already established. We have seen that (4) and (6) are true.[87] Proposition (5) follows from (3) and (4), while (7) follows from (5) and (6). Hence, (7) will be established if (3) is true. Since (3) follows from (1) and (2), it remains only to show that these must be accepted.

Aristotle himself articulates (1), and for good reason:

Non-synonymous things are all incommensurable.[88] Why is <it not possible to say> of what is not commensurable, e.g. a pen, the wine, or the highest note,

---

[87] See I. 3 and I. 8.

[88] Reading *mê synônuma, pant' asumblêta* with Ross at 248ᵇ7, as against the troubled manuscript variations. This reading seems required by 249ᵃ12–14. Another acceptable reading would be *homônuma panta asumblêta*, as in I, though this renders the discussion repetitious.

whether one is sharper <than the others>? The reason is that homonymous things are not commensurable. The highest note in a scale is, however, commensurable with the lead note, since 'sharp' signifies the same for both. (*Phys.* 248ᵇ6–11; cf. *Cat.* 11ᵃ5–13; *Pol.* 1259ᵇ36–8; *Top.* 116ᵃ1–8)

Unless two things are *F* synonymously, it is impossible to compare them in terms of *F*-ness.[89]

This is most obviously true in the case of discrete homonyms. If a desk and a problem in trigonometry are both hard, it is not possible to determine which one is harder. Similarly, if a novice and the grass are both green, it is not possible to determine which is greener. Of course, we may in poetic contexts draw such comparisons; but that is just because their literal inaptness jars us into superimposing one scale of comparison onto another. In these sorts of cases, literal comparison yields nonsense.

Equally, but less obviously, these considerations apply to associated homonyms as well. We call complexions and regimens healthy. If we try to determine which is healthier, we end up puzzled, since there is no standard of comparison. By contrast, it makes perfect sense to compare two regimens: one is healthier than the other because it is more productive of health. But neither regimen is more or less productive of health than a complexion is indicative of health.

Some have countered that Aristotle sometimes relaxed his general preference for refusing cross-categorial comparisons, and indeed that he was right to do so.[90] If some goods are instrumental for others, we may judge that the intrinsic goods are better than the instrumental, even though 'good' does not apply synonymously to both cases. Thus, if a healthy regimen is instrumental for health, we may say that health is better than a healthy regimen. In these cases, we evidently mean that health, considered as such, is preferable to a healthy regimen. But since these are paradigmatic homonyms, surely we can and do make cross-categorial comparisons among homonymous *F*s.

Caution is required here, however, since, as we have seen, homonymy is predicate-relative. That is, two things may be homonymously *F* but synonymously *G*. For example, an eye in a statue and a human eye may be

---

[89] Aristotle goes on to ask whether perhaps one might deny that non-homonyms are always synonymous (*Phys.* 248ᵇ12–21). This is to ask whether some synonyms may be incommensurable, not whether some homonyms turn out commensurable after all. Aristotle does not qualify the blanket statement of 248ᵇ8–9.

[90] Morrison (1987): 'Normally one is not allowed to compare across ambiguity. But when the items to which the ambiguous predicate is applied are related to each other as prior and posterior, then comparison is allowed.' The evidence is supposed to come from the *Protrepticus* 81–2; there, however, Aristotle seems to presuppose only that health is more choiceworthy than a healthy thing, where 'choiceworthy' will be applied synonymously. See also Pakuluk 1992.

homonymously eyes, even though they are synonymously magnitudes. Aristotle does not object to comparing two homonymous things in so far as they are also synonymous; he objects only in so far as they are homonymous. An eye in a statue may well be heavier than a human eye. Now, 'health' applies homonymously to regimens and persons; it should therefore be impossible to determine which is healthier. Even so, 'pleasant' may apply to them synonymously, as may 'desirable' or 'choiceworthy'. For each of these cases, it may then be possible to judge that health is worthier of being chosen than is a regimen, even though the regimen is itself worthy of choice.

This is something Aristotle himself makes perfectly clear in *Topics* iii. 1:

> Also, that which is choiceworthy through itself is more choiceworthy than what is choiceworthy through something else (*kai to di' hauto haireton tou di' heterou hairetou hairetôteron*). For example, health is more choiceworthy than gymnastics, since the one is choiceworthy through itself, and the other through something else. (*Top.* 116ª29–31)

These sorts of judgements do not contravene Aristotle's strictures against comparisons among homonyms, because the class of things being compared are synonymously choiceworthy,[91] if to different degrees.[92]

If we think that regimens and health are good things, then if 'good' is homonymous in this case, it should not be possible to say which is better. Where we have intuitions comparing these disparate sorts of things, it is along the lines of some single, synonymous predicate, like choiceworthiness. Presumably this is why Aristotle is careful in *Topics* iii. 1 and elsewhere to specify the predicate in terms of which things are to be judged better or worse, and why he himself follows his advice of comparing only synonymous *F*s (*Meteor.* 347ᵇ7; *HA* 608ᵇ5; *PA* 684ª25; *EN* 1120ª17–23; *Pol.* 1259ᵇ37).[93]

Hence, we see that Aristotle rightly accepts (1), the claim that two *F* things are non-synonymously *F* if and only if they are incommensurable as *F*s.

---

[91] This is not to claim that 'choiceworthy' (*haireton*) is synonymous. Rather, its applications in these cases are constrained to one sense. See *Topics* 118ᵇ27–37.

[92] Cf. *Topics* 117ᵇ10: where *kai* in *beltion kai hairetôteron* appears epexegetical; cf. as well *Topics* 119ª16–19.

[93] For some questions about Aristotle's practice, see Wardy 1990, 273–83. Wardy is wrong to think that Aristotle's treatment of *polu* at *Physics* 248ᵇ17–18 constrains him 'to believe that the vast majority of words are radically homonymous' (282). For the procedures there may show that there are many more cases of homonymy than we might have thought without showing that these cases are 'radical', where presumably this entails minimally that they are non-associated.

This leaves only (2), the claim that beings are always commensurable as beings. The principal argument for Aristotle's acceptance of (2) is his practice of treating different kinds of beings as commensurable. Thus, for example, in a striking passage from *Metaphysics* xiv. 1, Aristotle seeks to show why Platonists are wrong to treat contraries as first principles. In that context, he argues that relative terms are least of all substantial and least of all beings:[94]

[I]t is necessary that great and small, and whatever is of this sort are relative to something (*pros ti*). But the relative is least of all things some nature or some substance, and is posterior to quality and quantity . . . An indication that the relative is least of all some substance and some being is that of it alone there is no generation or destruction or movement, as in the case of quantity <where> there is increase and diminution, and in quality alteration, in place locomotion, and in substance generation and destruction *simpliciter*—but not for relatives. (*Met.* 1088ᵃ22–5, 29–34)

Here again Aristotle seems to claim that something can be less, or be less real, than other things. For he claims, quite arrestingly, that its not admitting of various kinds of changes indicates that the relative is 'least <of all> some substance and some being' (*hêkista ousia tis kai on ti to pros ti*; 1088ᵃ29–30; cf. *Cat.* 2ᵇ7–18, ᵇ22, 3ᵇ35–4ᵃ9, 15ᵃ4–7; *Met.* 1002ᵃ4–8, ᵃ15–18, 1029ᵃ6, ᵃ29–30, 1040ᵇ22–4, 1077ᵇ12). Of particular interest is the way Aristotle yokes substance (*ousia*) and being (*on*) here. Taken one way, he means merely 'some substance, that is, some being'; taken another way, he intends them conjunctively, simply 'some substance and some being'. On the plausible assumption that 'least of all' ([*pantôn*] *hêkista*)[95] ranges over both substance and being, Aristotle seems directly to assert that being can admit of degrees.[96] If that is so, then beings must be commensurable, since only what is commensurable admits of a more and a less. A crane in a construction site may be larger than a crane in a marsh; but it cannot be more or less a crane.

Simply put, if Aristotle accepts a degrees-of-reality hypothesis, then beings are commensurable. He asserts this thesis. This, then, suffices to establish that Aristotle endorses (2).

[94] Aristotle does not elsewhere treat relation as the most superficial of the categories. It may be that the list of categories considered at *Metaphysics* 1088ᵃ23–35 is abbreviated from the longer lists given at *Categories* 1ᵇ27 and *Topics* 103ᵇ23. See Ross 1924, ii. 473, and Apelt 1891, 140–1.

[95] Presumably Aristotle understands *pantôn hêkista*, as at 1088ᵃ23.

[96] Morrison (1987) argues that Aristotle in fact accepts a degrees-of-reality thesis. Frede and Patzig (1988) also seem to assume that he does: 'Wenn also die Form gegenüber der Materiel Priorität hat und in strengerem Sinne als diese ein Seiendes ist, so folgt, daß sie auch, aus demselben Grunde, Priorität haben muß gegenüber dem aus beiden Zusammengesetzten.'

Now, it is not clear that Aristotle should accept a degrees-of-reality thesis, since it is of dubious coherence.[97] Still, he is right to accept the commensurability of being.

To appreciate this, it is worth ending where we began, at the beginning of the *Categories*. There, recall, Aristotle asserts: 'Those things are called homonymous of which the name alone is common, but the account of being corresponding to the name is different . . . Those things are called synonymous of which the name is common, and the account of being corresponding to the name is the same' (1ª1–4, 6–7). This shows us that as a categorial doctrine we have the non-synonymy of 'existence' if and only if 'exist' receives different accounts in:

(1) Substances exist.

(2) Qualities exist.

(3) Quantities exist.

(4) Relatives exist.

And so on for the remaining categories of being. Now, it is worth stressing something regularly overlooked in discussions of the homonymy of being. The difference of account required for non-synonymy does not pertain to substances, qualities, quantities, or relatives. These do indeed have different accounts; but there is no question of their being synonymous as substances, qualities, quantities, or relatives (though they are synonymously categories). Rather, the difference of account pertains to 'exist', taken by itself. No (putative) facts about the ontic dependence relations among the various categories is relevant to the question of whether or not this account is univocal.[98] More important, in the present context, when we claim that substances and relatives exist, we are evidently saying something commensurable. When I say that there are relations but not points in absolute space, I am denying to one category of thing precisely what I am claiming on behalf of another.

[97] It would be most natural to restrict degrees to graded or norm kinds, or kinds which are functionally determined. Being does not seem to be such a kind. In response to the question 'Is *a* more than *b*?' we would normally be inclined to ask: 'Is *a* more *what* than *b*?' The degrees-of-reality thesis treats the initial question as coherent as it stands.

[98] So I disagree with Matthews (1995, 237), who gives an admirably clear expression to this intuition: 'When Aristotle makes the claim [that "is" is said in as many ways as there are categories], he generally adds that the use of "is" for substance is primary. What he seems to mean is that things in others categories exist if, and only if, they are the qualities, quantities, places, and so forth, *of substances*. In this way the existence of things in other categories is parasitic upon the existence of substances.' One can grant that qualities exist only if substances exist even while insisting that an account of 'exist' in 'substances exist' and 'qualities exist' will be synonymous.

Moreover, and in any case, there is no reason to believe even that the accounts of the various categories must make reference to substances. Relations can relate relations as their *relata* (e.g. it is more difficult to comprehend 'being a cosine of' than it is to comprehend 'being larger than'; here the relation 'being more difficult to comprehend' relates two relations). Similarly, quantities can number non-substances (the example of relations just mooted involves three relations). Qualities can be qualities of non-substances (e.g. blueness is a colour), and so forth. Hence, ontic dependence does not entail account dependence. And surely it is *account* dependence which is relevant to our establishing core-dependent homonymy. In any case, as we have seen, those who point to these forms of dependence already assume the non-univocity of being. Yet since cross-categorial existence claims are possible, claims regarding the *existence* of beings in different categories are as such commensurable.

Taking all this together: Aristotle sometimes asserts a degrees-of-reality thesis. This thesis presupposes that beings are commensurable. So he accepts a thesis which presupposes the truth of (2). Further, whatever its independent merits, the degrees-of-reality hypothesis appropriately assumes that beings as such are commensurable. For accounts of *existence* in claims of existence across the categories will be synonymous, no matter what ontic dependence relations obtain between members of the various categories.

This completes our challenge to the thesis that being is a core-dependent homonym. Since being as such is univocal, it cannot be a core-dependent homonym; for it is not a homonym at all.

9.10 CONCLUSIONS

Something seems initially plausible about Aristotle's contention that being is homonymous; indeed, something seems *deeply* plausible about this view. Yet when we attempt to defend some precise version of Aristotle's contention, we find it wanting. This is because the doctrine is false.

Its initial plausibility stems from its similarity to some other doctrines which are wholly defensible, e.g. that some beings exist necessarily and others contingently, or that some beings depend for their existence on the existence of other sorts of beings, or, indeed, that there are fundamentally different kinds of beings. None of these theses, however, states or entails the philosophically distinctive thesis Aristotle intends when introducing being as a core-dependent homonym. This thesis requires a separate defence, which Aristotle never provides. Moreover, the variety of defences his expositors have offered on his behalf fail. Indeed, that there is such an arresting expository variety already indicates some disarray in the doctrine: commentators have been talking about a host of different theses under the same label, with the result that importantly different theses are confused or partially conflated. Perhaps this tendency to restructure Aristotle's contentions in non-equivalent ways is not surprising. If the arguments I have mounted are cogent, there is no distinctively Aristotelian doctrine about the homonymy of being to illuminate or defend.

Afterword: Homonymy's Promise Reconsidered

The results of our investigation into Aristotle's approach to the homonymy of being have been mainly negative. If we agree that establishing core-dependent homonymy is a three-stage process, the first of which involves establishing non-univocity, then we are hard pressed to appreciate how Aristotle can show that being qualifies as core-dependent. It is open to him to argue, in a positive way, that non-substances depend for their existence on the existence of substances; but even winning this point will not suffice to establish the non-univocity of 'existence'. For it will be appropriate for the defender of synonymy to demand something more at this juncture: Aristotle must show that the accounts of 'exists' as it attaches to the various categories will diverge. This is something neither he nor his expositors have been able to accomplish.

Now, it would be unfortunate to allow this negative conclusion to overshadow the positive contribution that Aristotle's approach to homonymy has to offer. To begin, although I am sceptical that any Aristotelian will ever be able to establish the non-univocity of 'existence', I remain optimistic about Aristotle's contention that *many* central philosophical concepts are in fact core-dependent homonyms. The first point, then, is that it is possible to distinguish rather sharply between the *methodology* of homonymy, with its distinctive proposals about philosophical definition, and the claims its proponents offer regarding the range of its application. So, while 'being' fails to be homonymous, 'goodness', 'life', 'body', and 'oneness' all qualify. Further, although I have nowhere argued it in this book, I maintain that Aristotle's claims regarding a host of other homonyms are well-founded. These include: 'cause', 'principle', 'nature', 'necessity', 'substance', 'friendship', 'part', 'whole', 'priority', 'posteriority', 'the state', and 'justice'. To these, I would add, from some contemporary debates: 'mind', 'consciousness', 'law', 'responsibility', 'freedom', 'determination', 'property', 'right', 'concept', 'knowledge', 'consequence', and 'love'.

To take just one of these examples, which could be multiplied, it is worth reflecting on the notion of causation as it crops up in some contemporary discussions. Some theses regarding the nature of causation in contemporary debates include: (i) that a cause is a sufficient condition;[1]

---

[1] Burks 1951.

(ii) that a cause is a necessary condition;[2] (iii) that a cause is a necessary and sufficient condition;[3] (iv) that a cause is an INUS condition, that is, an insufficient part of an unnecessary but sufficient complex of conditions;[4] (v) that a cause is an event which raises the probability of another event's occurring to above 0.5;[5] (vi) a cause is whatever ensures that its effects are not coincidences;[6] (vii) a cause is a contingent connection of events falling under necessarily related universals;[7] or simply, (viii) a cause is a constant conjunction between event types.[8]

Judging from a sufficient remove, it is hard to escape the thought that each of these theories captures something important about causation; and it is tempting to infer on this basis, since they are in some cases incompatible with one another, that causation admits of no general univocal analysis. It is tempting to yield, that is, to the negative judgement that genuine philosophical analysis in this case is bound to come up empty-handed. This is in part, it seems, what has attracted some philosophers to Wittgensteinean family-resemblance hypotheses: while acknowledging some forms of overlap between the various cases of causation we encounter, they insist that there is nothing more to be said once their criss-crossing similarities are noted.

Aristotle rightly resists this temptation. Although, like some later theorists, he doubts that univocal definitions for central philosophical notions will be forthcoming, Aristotle nevertheless holds out for unity in multiplicity: he sees that although non-univocal, some philosophical concepts may nevertheless be ordered around a core in an asymmetric way. In the case of causation, he thinks that a cause is a kind of generative source of its result; but he doubts that this will amount to the same thing in every instance of causation.[9] In this case, as in others, Aristotle can fairly claim to have uncovered some logical space for analysis, a *tertium quid* between univocity, or definition in terms of necessary and sufficient conditions, and mere family resemblance. A philosophical account given in terms of core-dependent homonymy occupies just this space.

That some of Aristotle's attempted analyses should fail is hardly

---

[2] Lewis 1973.
[3] Taylor 1963.
[4] Mackie 1965.
[5] Mellor 1988.
[6] Owens 1992.
[7] Tooley 1987.
[8] All Humean accounts hold this.
[9] It is interesting to compare his attitude to that of Sosa (1993, 241), who comes close to Aristotle's view, even if he wants to hold out for a more general form of univocity. In addition to nomological causation, Sosa argues, there are material causation, consequentialist causation, and inclusive causation. These, he maintains, 'are distinct from one another and distinct also from the nomological causation that is parasitic on contingent general principles. At the same time these all seem to be types of causation in that most proper sense in which causes are sources of results or consequences that derive from them'.

surprising. Considering that he wanted to demonstrate that even the highest-level concepts could be subjected to analysis in terms of core-dependence, Aristotle may be guilty of a certain sort of overreaching. This does nothing to impugn his general method of analysis in terms of core-dependent homonymy; and it does nothing to undermine the successes he enjoyed with the method. Minimally, he is to be credited with showing that alternatives understood to be exhaustive are hardly so: for there may be multiplicity which the partisans of univocity miss and order in that multiplicity which their critics indefensibly disregard. Moreover, to the degree that his individual analyses in terms of core-dependent homonymy succeed, Aristotle can further be credited with making good on the promise his method offers. Indeed, one impressive legacy of his insight into core-dependent homonymy is the fruit his method continues to bear, even now, in a philosophical milieu in many ways alien to the one in which he found himself.

Bibliography

ACKRILL, J. L. (1963). *Categories and De Interpretatione* (Oxford).

—— (1972). 'Aristotle on "Good" and the Categories', in S. Stern, V. Brown, and A. Hourani, eds., *Islamic Philosophy and the Classical Tradition* (Oxford), 17–25.

—— (1972–3/1978). 'Aristotle's Definitions of *Psuchê*', *Proceedings of the Aristotelian Society*, 73: 119–33; reprinted in J. Barnes, M. Schofield, and R. Sorabji, eds., *Articles on Aristotle*, vol. iv (London, 1978), 65–75.

—— (1981). 'Aristotle's Theory of Definition: Some Questions on *Posterior Analytics* II. 8–10', in E. Berti, ed., *Aristotle on Science: The Posterior Analytics* (Padua), 359–84.

ALBRITTON, R. G. (1957). 'Forms of Particular Substances in Aristotle's *Metaphysics*', *Journal of Philosophy*, 54: 699–708.

ALLEN, D. J. (1936). *Aristotelis De Caelo* (Oxford).

ALLEN, R. E. (1973). 'Substance and Predication in Aristotle', in E. N. Lee, A. P. D. Mourelatos, and R. M. Rorty, eds., *Exegesis and Argument* (Assen), 362–73.

ALSTON, W. P. (1971). 'How Does one Tell if a Word has One, Several, or Many Senses?', in D. D. Steinberg and L. A. Jakobovits, eds., *Semantics* (Cambridge), 35–47.

ANSCOMBE, G. E. M. (1975). 'The Principle of Individuation', in J. Barnes, M. Schofield, and R. Sorabji, eds., *Articles on Aristotle*, 4 vols. (London, 1975-9), 88–95.

—— and GEACH, P. (1961). *Three Philosophers: Aristotle, Aquinas, Frege* (Oxford).

ANTON, J. (1968). 'The Aristotelian Doctrine of Homonuma in the *Categories* and its Platonic Antecedents', *Journal of the History of Philosophy*, 6: 315–26.

—— (1969). 'Ancient Interpretations of Aristotle's Doctrine of Homonuma', *Journal of the History of Philosophy*, 7: 1–18.

APELT, O. (1891). *Beitrage zur Geschichte der griechischen Philosophie* (Leipzig).

AQUINAS, T. (1950). *In Aristotelis Metaphysica*, ed. M. Cathala and R. Spiazzi (Turin).

ARENS, H. (1984). *Aristotle's Theory of Language and its Tradition* (Amsterdam).

ARMSTRONG, D. (1968). *A Materialist Conception of Mind* (London).

ARPE, C. (1938). *Das ti ên einai bei Aristoteles* (Hamburg).

ASHWORTH, E. J. (1995). 'Suárez on the Analogy of Being: Some Historical Background', *Vivarium*, 33: 50–75.

AUBENQUE, P. (1961). 'Sur la notion aristotelicienne de l'aporie', in S. Mansion, ed., *Aristote et les problèmes de methode* (Louvain), 3–19.

—— (1962). *Le Problème de l'être chez Aristote* (Paris).

—— (1978). 'Les Origines de la doctrine de l'analogie de l'être. Sur l'histoire d'un contresens', *Études philosophiques*, 49: 3–12.

—— (1987). 'Zur Entstehung der pseudoaristotelischen Lehre von der Analogie des Seins', in J. Wiesner, ed., *Aristoteles: Werk und Wirkung* (Berlin), ii. 233–48.

BAMBROUGH, J. R. (1960–1). 'Universals and Family Resemblances', *Proceedings of the Aristotelian Society*, 61: 207–22.

BARNES, J. (1970). 'Property in Aristotle's *Topics*', *Archiv für Geschichte der Philosophie*, 52: 136–55.

—— (1971). 'Homonymy in Aristotle and Speusippus', *Classical Quarterly*, 21: 65–80.

—— (1984). *Complete Works of Aristotle: The Revised Oxford Translation* (Princeton, NJ).

—— (1995). 'Metaphysics', *The Cambridge Companion to Aristotle* (Cambridge), 66–108.

—— SCHOFIELD, M., and SORABJI, R., eds. (1975–9). *Articles on Aristotle*, 4 vols. (London).

BARTH, T. (1942). 'Das Problem der Vieldeutigkeit bei Aristoteles', *Sophia*, 10: 11–30.

BEALER, G. (1982). *Quality and Concept* (Oxford).

BEDAU, M. (1996), 'The Nature of Life', in M. Boden, ed., *The Philosophy of Artificial Life* (Oxford), 332–57.

BLOCK, I. (1961). 'The Order of Aristotle's Psychological Writings', *American Journal of Philology*, 82: 50–77.

BLOCK, N. J. (1971). 'Are Mechanical and Teleological Explanations of Behaviour Incompatible?', *Philosophical Quarterly*, 21: 109–17.

BODEN, M. (1996). 'The Intellectual Context of Artificial Life', in M. Boden, ed., *The Philosophy of Artificial Life* (Oxford), 1–35.

BOLTON, R. (1976). 'Essentialism and Semantic Theory in Aristotle', *Philosophical Review*, 85: 514–44.

—— (1985). 'Aristotle on the Signification of Names', *Language and Reality in Greek Philosophy: Proceedings of the Greek Philosophical Society* (Athens), 153–62.

—— (1995). 'Science and the Science of Substance in Aristotle's *Metaphysics Z*', *Pacific Philosophical Quarterly*, 76: 419–69.

BONITZ, H. (1853). 'Ueber die Kategorien des Aristoteles', *Sitzungsberichte der Wiener Akademie*, 10: 591–645.

—— (1869). *Aristotelische Studien* (Hildesheim).

—— ed. (1848–9). *Aristotelis Metaphysica* (Bonn).

BOSTOCK, D. (1982). 'Aristotle on the Principles of Change in *Physics* i', in M. Schofield and M. Nussbaum, eds., *Language and Logos* (Cambridge), 179–96.

—— (1994). *Aristotle: Metaphysics Books Z and H* (Oxford).

BRENTANO, F. (1862/1975). *Von der mannigfachen Bedeutung des Seienden nach Aristoteles* (Freiburg im Breisgau). Available in English as *On the Several Senses of Being in Aristotle*, trans. R. George (Berkeley, Calif.).

BRUNSCHWIG, J. (1967). *Aristote: Topiques i–iv* (Paris).

—— (1979). 'La Forme, predicat de la matière?', in P. Aubenque, ed., *Études sur la Metaphysique d'Aristote* (Paris), 131–58.

BURKS, A. (1951). 'The Logic of Causal Propositions', *Mind*, 60: 263–82.

BURNET, J. (1900). *Aristotle: Ethics* (London).

BURNYEAT, M. (1992). 'Is an Aristotelian Philosophy of Mind Still Credible? (A Draft)', in M. Nussbaum and A. Rorty, eds., *Essays on Aristotle's De Anima* (Oxford), 15–26.

BUTCHVAROV, P. (1982). 'That Simple, Indefinable, Nonnatural Property Good', *Review of Metaphysics*, 36: 51–75.

CARNAP, R. (1956). *Meaning and Necessity* (Chicago).

CASTAÑEDA, H. (1972). 'Thinking and the Structure of the World', *Critica*, 6: 43–86.

CHARLES, D. (1994). 'Aristotle on Names and their Signification', in S. Everson, ed., *Language* (Cambridge).

CHARLTON, W. (1970). *Aristotle: Physics I, II* (Oxford).

—— (1972). 'Aristotle and the Principle of Individuation', *Phronesis*, 17: 239–49.

—— (1994). 'Aristotle on Identity', in T. Scaltsas, D. Charles, and M. L. Gill, eds., *Unity, Identity and Explanation in Aristotle's Metaphysics* (Oxford), 41–53.

CHERNISS, H. (1935). *Aristotle's Criticism of Presocratic Philosophy* (Baltimore, Md.).

—— (1944). *Aristotle's Criticism of Plato and the Academy* (Baltimore, Md.).

CLEARY, J. (1988). *Aristotle on the Many Senses of Priority* (Carbondale, Ill.).

CODE, A. (1976). 'The Persistence of Aristotelian Matter', *Philosophical Studies*, 29: 357–67.

—— (1978). 'No Universal is a Substance', *Paideia*, 65–74.

—— (1984). 'The Aporematic Approach to Primary Being', *Canadian Journal of Philosophy*, Supp. 10: 41–65.

—— (1985). 'On the Origins of some Aristotelian Theses about Predication', in J. Bogan and J. McGuire, eds., *How Things Are: Studies in Predication and the History of Philosophy* (Doredrecht).

—— (1986). 'Aristotle's Investigation of a Basic Logical Principle', *Canadian Journal of Philosophy*, 16: 341–57.

—— (n.d.). 'Aristotle's Metaphysics as a Science of Principles' (unpublished type-script).

COHEN, S. M. (1987). 'The Credibility of Aristotle's Philosophy of Mind', in M. Matthen, ed., *Aristotle Today* (Edmonton, Alberta).

COOPER, J. M. (1977). 'Aristotle on the Forms of Friendship', *Review of Metaphysics*, 30: 619–48.

—— (1985). 'Aristotle on the Goods of Fortune', *Philosophical Review*, 94: 173–96.

COPI, I. (1954). 'Essence and Accident', *Journal of Philosophy*, 51: 706–19.

DANCY, R. (1975). 'On Some of Aristotle's First Thoughts about Substance', *Philosophical Review*, 84: 338–73.

—— (1978). 'On Some of Aristotle's Second Thoughts about Substances: Matter', *Philosophical Review*, 87: 372–413.

DEVEREUX, D. T. (1985). 'The Primacy of *Ousia*', in D. J. O'Meara, ed., *Platonic Investigations* (Washington, DC), 219–46.

DRETSKE, F. (1995). *Naturalizing the Mind* (Cambridge, Mass.).

DRISCOLL, J. (1981). '*Eidê* in Aristotle's Earlier and Later Theories of Substance', in D. J. O'Meara, ed., *Studies in Aristotle* (Washington, DC), 129–59.

DUERLINGER, J. (1970). 'Predication and Inherence in Aristotle's *Categories*', *Phronesis*, 15: 179–203.

DUMMETT, M. (1975). 'What is a Theory of Meaning?' in S. D. Guttenplan, ed., *Mind and Language* (Oxford).

DURING, I. and OWEN, G. E. L., eds. (1960). *Aristotle and Plato in the Mid-Fourth Century* (Gothenburg).

EVANS, J. D. G. (1987). *Aristotle* (New York).

FEREJOHN, M. (1980). 'Aristotle on Focal Meaning and the Unity of Science', *Phronesis*, 25: 117–28.

FINE, G. (1982). 'Owen, Aristotle, and the Third Man', *Phronesis*, 27: 13–33.

—— (1983). 'Plato and Aristotle on Form and Substance', *Proceedings of the Cambridge Philological Society*, 209: 23–47.

—— (1984). 'Separation', *Oxford Studies in Ancient Philosophy*, 2: 31–87.

—— (1988). 'Owen's Progress', *Philosophical Review*, 97: 373–99.

—— (1993). *On Ideas: Aristotle's Criticism of Plato's Theory of Forms* (Oxford).

FINE, K. (1994). 'A Puzzle Concerning Matter and Form', in T. Scaltsas, D. Charles, and M. L. Gill, eds., *Unity, Identity, and Explanation in Aristotle's Metaphysics* (Oxford), 13–40.

FORTENBAUGH, W. W. (1975). 'Aristotle's Analysis of Friendship', *Phronesis*, 20: 51–62.

FREDE, M. (1987). 'The Unity of Special and General Metaphysics', *Essays in Ancient Philosophy* (Oxford), 81–95.

—— and PATZIG, G. (1988). *Aristoteles: Metaphysik Z* (Munich).

FURTH, M. (1988) *Substance, Form, and Psyche: An Aristotelian Metaphysics* (Cambridge).

GAISER, K. (1985). *Theophrast in Assos: Zur Entwicklung der Naturalwissenschaft zwischen Akademie und Peripatos* (Heidelberg).

GAUTHIER, R. A., and JOLIF, J. Y., trans. and eds. (1970). *Aristote: l'Ethicque à Nicomaque*, 4. vols. (Louvain).

GEACH, P. T. (1954–5). 'Form and Existence', *Proceedings of the Aristotelian Society*, 55: 251–72.

GEYSER, J. (1917). *Die Erkenntnistheorie des Aristoteles* (Münster).

GILL, M. L. (1989). *Aristotle on Substance: The Paradox of Unity* (Princeton, NJ).

—— (1994). 'Individuals and Individuation in Aristotle', in T. Scaltsas, D. Charles, and M. L. Gill, eds., *Unity, Identity, and Explanation in Aristotle's Metaphysics* (Oxford).

GOTTHELF, A., ed. (1985). *Aristotle on Nature and Living Things* (Pittsburgh, Pa.).

GRAHAM, D. (1987). *Aristotle's Two Systems* (Oxford).

GRAHAM, W. (1975). 'Counterpredictability and *Per Se* Accidents', *Archiv für Geschichte der Philosophie*, 57: 182–7.

GRANGER, H. (1981). 'The Differentia and the *Per Se* Accident in Aristotle', *Archiv für Geschichte der Philosophie*, 63: 118–29.

—— (1984). 'Aristotle on Genus and Differentiae', *Journal of the History of Philosophy*, 22: 1–23.

GRICE, P. (1988). 'Aristotle on the Multivocity of Being', *Pacific Philosophical Quarterly*, 69: 175–200.

HALLER, R. (1962). 'Untersuchungen zur Bedeutungsproblem in der antiken und mittelalterlichen Philosophie', *Archiv für Begriffsgeschichte*, 7: 57–119.

HAMBRUCH, E. (1904). *Logische Regeln der Platonischen Schule in der Aristotelischen Topik* (Berlin).

HAMLYN, D. W. (1977–8). 'Focal Meaning', *Proceedings of the Aristotelian Society*, 78: 1–18.

HARDIE, W. F. R. (1980). *Aristotle's Ethical Theory* (Oxford).

HARTMAN, E. (1976). 'Aristotle on the Identity of Substance and Essence', *Philosophical Review*, 85: 545–61.

—— (1977). *Substance, Body, and Soul* (Princeton, NJ).

HEINAMAN, R. E. (1979). 'Aristotle's Tenth Aporia', *Archiv für Geschichte der Philosophie*, 61: 249–70.

—— (1982). 'Form and Universal in Aristotle', *Classical Review*, 31: 44–8.

HICKS, R. D. (1907). *Aristotle: De Anima* (Cambridge).

HINTIKKA, K. J. (1971). 'Different Kinds of Equivocation in Aristotle', *Journal of the History of Philosophy*, 9: 368–72.

—— (1973). 'Aristotle and the Ambiguity of Ambiguity', in K. J. Hintikka, ed., *Time and Necessity* (Oxford), ch. 1.

HIRSCH, E. (1982). *The Concept of Identity* (Oxford).

IRWIN, T. H. (1980). 'The Metaphysical and Psychological Basis of Aristotle's Ethics', in A. Rorty, ed., *Essays on Aristotle's Ethics* (Berkeley, Calif.), 35–54.

—— (1981). 'Homonymy in Aristotle', *Review of Metaphysics*, 34: 523–44.

—— (1982). 'Aristotle's Concept of Signification', in M. Schofield and M. Nussbaum, eds., *Language and Logos* (Cambridge), 241–66.

—— (1986). *Aristotle's Ethica Nicomachea* (Indianapolis, Ind.).

—— (1988). *Aristotle's First Principles* (Oxford).

JAEGER, W. W. (1912). *Studien zur Entstehungsgeschichte der Metaphysik des Aristoteles* (Berlin).

—— (1948). *Aristotle: Fundamentals of the History of his Development* (Oxford).

—— ed. (1957). *Aristotelis Metaphysica* (Oxford).

JOACHIM, H. H. (1922). *Aristotle on Coming-to-be and Passing-away* (Oxford).

—— (1951). *Aristotle: Nicomachean Ethics* (Oxford).

JOSEPH, H. (1906). *An Introduction to Logic* (Oxford).

KATZ, J. (1990). *The Metaphysics of Meaning* (Cambridge, Mass.).

—— (1994). 'Precis of *The Metaphysics of Meaning*', *Philosophy and Phenomenological Research*, 54: 127–32.

KIRWAN, C. A. (1970–1). 'How Strong Are the Objections to Essence?', *Proceedings of the Aristotelian Society*, 71: 43–59.

—— (1971). *Aristotle: Metaphysics IV, V, VI* (Oxford).

KITCHER, P. (1993). 'Function and Design', in T. French, Jr., and H. Wettstein, eds., *Midwest Studies in Philosophy* (Notre Dame, Ind.).

KNEALE, W. C., and KNEALE, M. (1962). *The Development of Logic* (Oxford).

KOSMAN, A. (1968). 'Predicating the Good', *Phronesis*, 13: 171–4.

KRAUT, R. (1979). 'The Peculiar Function of Human Beings', *Canadian Journal of Philosophy*, 9: 467–78.
—— (1989). *Aristotle on the Human Good* (Princeton, NJ).
—— (1995). 'Review of Fine', *Philosophical Review*, 104: 116–18.
KRETZMANN, N. (1974). 'Aristotle on Spoken Sound Significant by Convention', in J. Corcoron, ed., *Ancient Logic and Its Modern Interpretations* (Dordrecht), 3–21.
KUNG, J. (1977). 'Aristotle on Essence and Explanation', *Philosophical Studies*, 31: 361–83.
LAPORTE, J. (1996). 'Chemical Kind Term Reference and the Discovery of Essence', *Noûs*, 30: 112–32.
LESZL, W. (1970). *Logic and Metaphysics in Aristotle* (Padua).
—— (1975). *Aristotle's Conception of Ontology* (Padua).
LEWIS, D. (1973). 'Causation', *Journal of Philosophy*, 70: 556–67.
—— (1986). *On the Plurality of Worlds* (Oxford).
LEWIS, F. (1982). 'Accidental Sameness in Aristotle', *Philosophical Studies*, 42: 1–36.
—— (1984). 'What is Aristotle's Theory of Essence?', *Canadian Journal of Philosophy*, Supp. 10: 89–131.
—— (1985). 'Form and Predication in Aristotle's Metaphysics', in J. Bogan and E. J. McGuire, eds., *How Things Are* (Dordrecht), 59–83.
—— (1991). *Substance and Predicate in Aristotle* (New York).
—— (1994). 'Aristotle on the Relation between a Thing and its Matter', in T. Scaltsas, D. Charles, and M. L. Gill, eds., *Unity, Identity, and Explanation in Aristotle's Metaphysics* (Oxford), 247–77.
LOUX, M. J. (1973). 'Aristotle on the Transcendentals', *Phronesis*, 18: 225–39.
—— (1979). 'Form, Species and Predication in *Metaphysics Zeta, Eta*, and *Theta*', *Mind*, 88: 1–23.
LUDWIG, K. (1994). 'Causal Relevance and Thought Content', *Philosophical Quarterly*, 44: 334–52.
MACDONALD, S. (1989). 'Aristotle and the Homonymy of the Good', *Archiv für Geschichte der Philosophie*, 71: 150–74.
MACKIE, J. (1965). 'Causes and Conditions', *American Philosophical Quarterly*, 2: 245–64.
MANSION, A. (1958). 'Philosophie première, philosophie seconde et metaphysique chez Aristote', *Revue philosophique de Louvain*, 56: 165–221.
MANSION, S. (1955). 'Les Apories de la metaphysique aristotelicienne', *Autour d'Aristote* (Louvain), 141–79.
—— (1976). *Le Jugement d'existence chez Aristote* (Louvain).
—— ed. (1961). *Aristote et les problèmes de methode* (Louvain).
MATTHEWS, G. B. (1972). 'Senses and Kinds', *Journal of Philosophy*, 69: 149–57.
—— (1977). 'Consciousness and Life', *Philosophy*, 52: 13–26.
—— (1982). 'Accidental Unities', in M. Schofield and M. Nussbaum, eds., *Language and Logos* (Cambridge), 223–40.
—— (1992). '*De Anima* 2. 2–4 and the Meaning of Life', in M. Nussbaum and A. Rorty, eds., *Essays on Aristotle's De Anima* (Oxford), 185–94.

—— (1995). 'Aristotle on Existence', *Bulletin of the Institute of Classical Studies*, 233–8.

MCINERNY, R. (1992). 'Aquinas and Analogy: Where Cajetan Went Wrong', *Philosophical Topics*, 20: 103–24.

MELLOR, D. H. (1988). 'On Raising the Chances of Some Effects', in J. Fetzer, ed., *Probability and Causation*, (Dordrecht), 229–39.

MILLER, F. D. (1973). 'Did Aristotle Have a Concept of Identity?', *Philosophical Review*, 82: 483–90.

—— (1995). *Nature, Justice and Rights in Aristotle's Politics* (Oxford).

MINIO-PALUELLO, L., ed. (1956). *Aristotelis Categoriae et Liber De Interpretatione* (Oxford).

—— ed. (1964). *Aristotelis Analytica Priora et Posteriora* (Oxford).

MODRAK, D. K. (1979). 'Forms, Types, and Tokens in Aristotle's *Metaphysics*', *Journal of Philosophy*, 17: 371–81.

—— (1987). *Aristotle: The Power of Perception* (Chicago, Ill.).

MORAVCSIK, J. M. E. (1967). 'Aristotle on Predication', *Philosophical Review*, 76: 80–96.

—— (1975). '*Aitia* as Generative Factor in Aristotle's Philosophy', *Dialogue*, 14: 622–38.

MORRISON, D. (1987). 'Evidence for Degrees of Being in Aristotle', *Classical Quarterly*, 37: 382–401.

NEANDER, K. (1991). 'The Teleological Notion of Function', *Australasian Journal of Philosophy*, 454–68.

NUCHELMANS, G. (1973). *Theories of the Proposition: Ancient and Medieval Conceptions of the Bearers of Truth and Falsity* (Amsterdam).

NUSSBAUM, M. (1982). 'Saving Aristotle's Appearances', in M. Schofield and M. Nussbaum, eds., *Language and Logos* (Cambridge).

—— (1986). *The Fragility of Goodness* (Cambridge).

—— trans. and ed. (1978). *Aristotle: De Motu Animalium* (Princeton, NJ).

—— and PUTNAM, H. (1992). 'Changing Aristotle's Mind', in M. Nussbaum and A. Rorty, eds., *Essays on Aristotle's De Anima* (Oxford), 27–56.

NUYENS, F. J. (1948). *L'Évolution de la psychologie d'Aristote* (Louvain).

OWEN, G. E. L. (1957). 'A Proof in the *Peri Ideôn*', *Journal of Hellenic Studies*, 77: 103–11; reprinted in M. Nussbaum, ed., *Logic, Science and Dialectic* (Ithaca, NY, 1986).

—— (1960). 'Logic and Metaphysics in Some Earlier Works of Aristotle', in I. During and G. E. L. Owen, eds., *Aristotle and Plato in the Mid-Fourth Century* (Goteborg), 163–90; reprinted in M. Nussbaum, ed., *Logic, Science and Dialectic* (Ithaca, NY, 1986).

—— (1961). 'Tithenai ta Phainomena', in S. Mansion, ed., *Aristote et les Problèmes de la Méthode* (Louvain), 83–103; reprinted in M. Nussbaum, ed., *Logic, Science and Dialectic* (Ithaca, NY, 1986).

—— (1965). 'Aristotle on the Snares of Ontology', in J. R. Bambrough, ed., *New Essays on Plato and Aristotle* (London), 69–75; reprinted in M. Nussbaum, ed., *Logic, Science and Dialectic* (Ithaca, NY, 1986).

OWEN, G. E. L. (1965). 'The Platonism of Aristotle', *Proceedings of the British Academy*, 50: 125–50; reprinted in M. Nussbaum, ed., *Logic, Science and Dialectic* (Ithaca, NY, 1986).

OWENS, D. (1992). *Causes and Coincidences* (Cambridge).

OWENS, J. (1951). *The Doctrine of Being in the Aristotelian Metaphysics* (Toronto, Ont.).

PAKULUK, M. (1992). 'Friendship and the Comparison of Goods', *Phronesis*, 37: 111–30.

PATZIG, G. (1960). 'Theologie und Ontologie in der *Metaphysik* des Aristoteles', *Kant-studien*, 52: 185–205; trans. as 'Theology and Ontology in Aristotle's *Metaphysics*', in J. Barnes, M. Schofield, and R. Sorabji, eds., *Articles on Aristotle*, 4 vols. (London, 1975–9), 33–49.

PELLETIER, F. (1979). 'Sameness and Referential Opacity in Aristotle', *Noûs*, 13: 283–311.

POSTE, E. (1866). *Aristotle on Fallacies; or, The Sophistici Elenchi* (London).

PUTNAM, H. (1964). 'Robots: Machines or Artificially Created Life?', *Journal of Philosophy*, 61: 668–91; reprinted in H. Putnam, *Mind, Language and Reality: Philosophical Papers*, vol. ii (Cambridge, 1975), 386–407.

—— (1975). ' "The Meaning of "Meaning" ' ', *Mind, Language and Reality: Philosophical Papers*, vol. ii (Cambridge), ch. 12.

—— (1975). 'On Properties', *Mind, Language and Reality: Philosophical Papers*, vol. ii (Cambridge), ch. 19.

—— (1981). *Reason, Truth, and History* (Cambridge).

RAPP, C. (1992). 'Ähnlichkeit, Analogie und Homonymie bei Aristoteles', *Zeitschrift für Philosophische Forschung*, 46: 526–44.

ROBINSON, H. (1978). 'Mind and Body in Aristotle', *Classical Quarterly*, 28: 107–24.

ROSS, G. R. T. (1906). *De Sensu et De Memoria* (Cambridge).

ROSS, W. D. (1923). *Aristotle* (London).

—— (1924). *Aristotle: Metaphysics* (Oxford).

—— (1936). *Aristotle: Physics* (Oxford).

—— (1949). *Aristotle: Prior and Posterior Analytics* (Oxford).

—— (1950). *Aristotelis Physica* (Oxford).

—— (1955*a*). *Aristotelis Fragmenta Selecta* (Oxford).

—— (1955*b*). *Aristotle: Parva Naturalia* (Oxford).

—— (1957). 'The Development of Aristotle's Thought', *Proceedings of the British Academy*, 43: 63–78; reprinted in J. Barnes, M. Schofield, and R. Sorabji, eds., *Articles on Aristotle*, 4 vols. (London, 1975–9), 1–13.

—— (1958). *Aristotelis Topic et Sophistici Elenchi* (Oxford).

RYLE, G. (1951). 'Thinking and Language', *Proceedings of the Aristotelian Society*, Supp. 25: 65–82.

SCALTSAS, T. (1994). 'Individuals and Individuation in Aristotle', in T. Scalstas, D. Charles, and M. L. Gill, eds., *Unity, Identity, and Explanation in Aristotle's Metaphysics* (Oxford): 247–77.

SCHIFFER, S. (1987). *Remnants of Meaning* (Cambridge, Mass.).

SCHOFIELD, M. (1975). 'Aristotle on the Imagination', in J. Barnes, M. Schofield, and R. Sorabji, eds., *Articles on Aristotle*, 4 vols. (London, 1975–9), 102–32.

SELLARS, W. S. (1957). 'Substance and Form in Aristotle', *Journal of Philosophy*, 54: 688–99.

SHIELDS, C. (1987). 'Commentary on Kosman's "Divine Being and Divine Thinking in *Metaphysics* Lambda"', *Proceedings of the Boston Area Colloquium on Ancient Philosophy*, 3: 189–201.

—— (1988a). 'Soul and Body in Aristotle', *Oxford Studies in Ancient Philosophy*, 6: 103–37.

—— (1988b). 'Soul as Subject in Aristotle's *De Anima*', *Classical Quarterly*, n.s., 38: 140–9.

—— (1990). 'The First Functionalist', in J.-C. Smith, ed., *Essays on the Historical Foundations of Cognitive Science* (Dordrecht), 19–33.

—— (1995). 'Intentionality and Isomorphism in Aristotle', in *Proceedings of the Boston Area Colloquium in Ancient Philosophy*, 11: 307–30.

SOLMSEN, F. (1929). *Die Entwicklung der aristotelischen Logik und Rhetorik* (Berlin).

SORABJI, R. (1971). 'Aristotle on Demarcating the Five Senses', *Philosophical Review*, 80: 55–79; reprinted in J. Barnes, M. Schofield, and R. Sorabji, eds., *Articles on Aristotle*, vol. iv (London, 1978), 76–92.

—— (1974). 'Body and Soul in Aristotle', *Philosophy*, 49: 63–89; reprinted in J. Barnes, M. Schofield, and R. Sorabji, eds., *Articles on Aristotle*, vol. iv (London, 1978), 42–64.

—— (1980). *Necessity, Cause, and Blame* (London).

SOSA, E. (1993). 'Varieties of Causation', in E. Sosa and M. Tooley, eds., *Causation* (Oxford), 234–42.

SPELLMAN, L. (1995). *Substance and Separation in Aristotle* (Cambridge).

STOUGH, C. L. (1972). 'Language and Ontology in Aristotle's *Categories*', *Journal of the History of Philosophy*, 10: 261–72.

STRAWSON, P. (1987). 'Concepts and Properties or Predication and Copulation', *Philosophical Quarterly*, 37: 402–6.

TARÁN, L. (1978). 'Speusippus and Aristotle on Homonymy and Synonymy', *Classical Quarterly*, 21: 65–80.

TAYLOR, R. (1963). 'Causation', *The Monist*, 47: 287–313.

TOOLEY, M. (1987). *Causation: A Realist Approach* (Oxford).

TRENDELENBURG, F. A. (1828). 'Das *to heni einai, to agathô(i) einai* etc., und das *to ti ên einai* bei Aristoteles: Ein Beitrag sur Aristotelischen Begriffsbestimmung und zur Greichischen Syntax', *Rheinisches Museum für Philologie*, 2: 457–83.

—— (1846). *Geschichte der Kategorienlehre* (Berlin).

URMSON, J. O. (1970). 'Polymorphous Concepts', in O. Wood and G. Pitcher, eds., *Ryle* (London), 249–68.

—— (1987). *Aristotle's Ethics* (Oxford).

VON WRIGHT, H. (1963). *The Varieties of Goodness* (London).

WAGNER, H. (1961). 'Über das Aristotelische *pollachôs legetai to on*', *Kant-studien*, 53: 75–91.

WAITZ, T. (1844–6). *Aristotelis Organon Graece* (Leipzig).
WALKER, M. (1979). 'Aristotle's Account of Friendship in the Nicomachean Ethics', *Phronesis*, 24: 180–96.
WARDY, R. (1990). *The Chain of Change* (Cambridge).
WATERLOW, S. (1982). *Nature, Change, and Agency in Aristotle's Physics* (Oxford).
WEDIN, M. (1988). *Mind and Imagination in Aristotle* (New Haven, Conn.).
—— (1991). 'Partisanship in *Metaphysics Z*', *Ancient Philosophy*, 11: 361–85.
WHITE, N. P. (1971). 'Aristotle on Sameness and Oneness', *Philosophical Review*, 80: 177–97.
WHITING, J. E. (1986). 'Form and Individuation in Aristotle', *History of Philosophy Quarterly*, 3: 359–77.
—— (1988). 'Aristotle's Function Argument: A Defence', *Ancient Philosophy*, 8: 33–48.
—— (1992). 'Living Bodies', in M. Nussbaum and A. Rorty, eds., *Essays on Aristotle's De Anima* (Oxford), 75–91.
WIELAND, W. (1962). *Die aristotelische Physik* (Göttingen).
—— (1970). *Die aristotelische Physik* (Göttingen).
WIGGINS, D. (1967). *Identity and Spatio-Temporal Continuity* (Oxford).
—— (1971). 'On Sentence-Sense, Word-Sense, and Difference of Word-Sense: Toward a Philosphical Theory of Dictionaries', in D. D. Steinberg and L. A. Jakobovits, eds., *Semantics* (Cambridge), 14–34.
—— (1980). *Sameness and Substance* (Cambridge, Mass.).
—— (1984). 'The Sense and Reference of Predicates: A Running Repair to Frege's Doctrine and a Plea for the Copula', *Philosophical Quarterly*, 34: 311–28.
WILLIAMS, B. (1986). 'Hylomorphism', *Oxford Studies in Ancient Philosophy*, 4: 186–99.
WILLIAMS, C. J. F. (1982). *Aristotle: De Generatione et Corruptione* (Oxford).
—— (1985). 'Aristotle's Theory of Descriptions', *Philosophical Review*, 94: 63–80.
WITTGENSTEIN, L. (1958). *Philosophical Investigations* (Oxford).
WOODFIELD, A. (1976). *Teleology* (Cambridge).
WOODS, M. J. (1965). 'Identity and Individuation', in R. J. Butler, ed., *Analytical Philosophy (Second Series)* (Oxford), 120–30.
—— (1967). 'Problems in *Metaphysics Z* 13', in J. Moravcsik, ed., *Aristotle: A Collection of Critical Essays* (Garden City, NY), 215–38.
—— (1974–5). 'Substance and Essence in Aristotle', *Proceedings of the Aristotelian Society*, 75: 167–80.
—— trans. and ed. (1982). *Aristotle: Eudemian Ethics i, ii, viii* (Oxford).
WRIGHT, L. (1976). *Teleological Explanations* (Berkeley, Calif.).
YABLO, S. (1987). 'Identity, Essence, and Indiscernibility', *Journal of Philosophy*, 84: 292–314.
ZALTA, E. (1988). *Intensional Logic and the Metaphysics of Intentionality* (Cambridge, Mass.).
ZIFF, P. (1960). *Semantic Analysis* (Ithaca, NY).

Index of Passages Cited

General Index